Green Energy and Technology

S.M. Muyeen • Junji Tamura • Toshiaki Murata

Stability Augmentation of a Grid-connected Wind Farm

Springer

Dr. S.M. Muyeen, JSPS (Japan Society for the Promotion of Science) Postdoctoral Fellow
Dr. Junji Tamura, Professor
Dr. Toshiaki Murata, Associate Professor

Kitami Institute of Technology
165 Koen Cho, Kitami
Hokkaido, 090-8507
Japan

ISBN 978-1-84800-315-6 e-ISBN 978-1-84800-316-3

DOI 10.1007/978-1-84800-316-3

Green Energy and Technology ISSN 1865-3529

British Library Cataloguing in Publication Data
Muyeen, S. M.
Stability augmentation of a grid-connected wind farm. -
(Green energy and technology)
1. Wind power plants 2. Wind turbines - Mathematical models
3. Wind energy conversion systems - Stability
4. Energy storage 5. Hydrogen as fuel
I. Title II. Tamura, Junji III. Murata, Toshiaki
621.3'12136
ISBN-13: 9781848003156

Library of Congress Control Number: 2008936056

Cover design: WMXDesign, Heidelberg, Germany

Printed on acid-free paper

9 8 7 6 5 4 3 2 1

springer.com

Foreword

Today, wind power is no longer "alternative". From being a token marginal afterthought, wind power has become a significant mainstream generating technology for electricity systems in many power systems around the world.

In Denmark, for instance, wind power supplies more than 20% of the local consumption and the Danish government aims to increase this share to 50% by 2025. The EU wide target is set at a penetration level of 20% by 2020. Hence, many countries around the world see no alternative to renewable energy in general and wind power in particular.

The implementation of these ambitious goals will lead to many challenges, in particular to the integration of wind power into the power system. When today's power systems were designed, large amounts of fluctuating power sources such as wind power were not an issue. Hence, increasing wind power penetration levels will require a gradual redesign of the existing power systems and operating approaches. This will include new planning approaches with detailed modeling and simulation of the impact that wind power has on the power system. Different solutions are already being developed in several countries depending on various factors such as power system structure, national and international regulatory framework and wind power penetration levels. In Western Denmark, for instance, there are already today instances when wind power supplies more than 100% of local consumption, which currently is possible only due to strong interconnections to neighboring countries.

While power systems certainly have to adapt to wind power, e.g., by becoming more flexible, wind power also has to adapt to the need of power systems. The development of grid codes for wind farms around the world is a first important step towards this adaptation. Fault-ride-through requirements for wind turbines/wind farms are an important technical feature of wind farms. This grid code requirement addresses an important issue regarding the large-scale integration of wind power. Additional important grid code features are ramp rate limitations and dynamic voltage support. As today grid codes require wind farms to meet the same standards regarding grid support as conventional power plants, such installations deserve to be called wind power plants rather than wind farms.

The next important step in this development is the integration of wind power plants into power system state estimators in combination with remote control of wind power plants, e.g., including ramping and power factor control, as well as frequency control support by wind farms. In addition the inertia contribution of wind power plants is discussed in some countries. This can be achieved by including a special control system in a wind turbine that mirrors an inertia-like behavior commonly found in conventional power plants.

All in all, there is an increasing need for wind power plants to adapt to the needs of the power system. Therefore, the topic of this book *Stability Augmentation of a Grid-connected Wind Farm* is very timely and provides an important contribution to the increasing use of wind power by large wind power plants in Japan and around the world.

Thomas Ackermann
Energynautics GmbH, Langen, Germany
Royal Institute of Technology, Stockholm, Sweden

Preface

This book is written in a somewhat tutorial style suitable for engineering students and professionals who are working with wind energy. It is also intended to be used by anyone with a good background in mathematics and physics who wants to be an expert in the field of wind energy.

This is a comprehensive approach to stabilizing a wind farm that may consist of fixed or variable speed wind turbine generator systems. This book provides advanced technical depth on wind turbine generator systems considering both mechanical and electrical sections. In the mechanical section, wind turbine drive train modeling and pitch control are emphasized. On the other hand, the electrical section is enriched with different types of facts controllers that can be adopted at a wind farm terminal.

Most of the chapters of the book are centered on a particular tool applicable to a wind farm. Besides the general discussion on the tool, detailed modeling and a control strategy for that system are discussed. The chapters are not limited to modeling and control systems of different types of tools, but incorporate extensive simulation results that will be very helpful to students. For the simulation analysis, the most popular digital simulator for power system, named PSCAD/EMTDC is used.

Chapter 1 consists of a general discussion on the recent status of wind power worldwide and some recent technological overviews on wind turbine generator systems. Chapter 2 discusses the modeling of both fixed and variable speed wind turbine including the drive train. Chapter 3 is focused on the design and control of different types of pitch controllers. Power smoothing of a wind turbine generator system by using pitch controller is a salient feature of this chapter. In Chap. 4, the STATCOM is emphasized for use with a fixed speed wind farm. Chapter 5, the heart of this book, focuses on different types of energy storage systems suitable for wind power application. In Chap. 6, hydrogen generation using wind power is described. Chapter 7 is related to both Chapters 5 and 6. A new wind farm operating strategy integrating an energy storage system and a hydrogen generator is presented, including detailed modeling and a control strategy. Wind power and terminal voltage smoothing of a wind farm are the salient features of this chapter. In Chap. 8, the stability analysis of variable speed wind turbine driving a permanent magnet synchronous generator is emphasized. The Appendix and the reference list are shown in Chap. 9 and Chap. 10 respectively.

In this book, one may find the essence of fixed and variable speed wind turbine generator systems, pitch control, hydrogen generation, etc. Significant technical depth can also be obtained on the energy storage systems (ESS) such as superconducting magnetic energy storage (SMES) system, energy capacitor system (ECS), flywheel energy storage system (FESS), and STATCOM/BESS, which can be integrated at the wind farm terminal.

Acknowledgement

A large number of individuals and organizations have assisted the authors in a variety of ways in the preparation of this work. In particular, we would like to thank Prof. Naoto Kakimoto, Prof. Hiroshi Tanimoto, Dr. Takao Ueda, and Dr. Mohd. Hasan Ali for their tremendous support and their peer review of each chapter of the book.

We have made extensive use of publications of the European Wind Energy Association (EWEA), the Global Wind Energy Council (GWEC), the American Wind Energy Association (AWEA), the Sandia National Laboratories, and the National Renewable Energy Laboratory (NREL) and record our special thanks to these organizations for making documents available to us free of charge and for permission to use some of the material therein.

The authors are grateful to ENERCON GmbH, REpower Systems AG, ABB, and Beacon Power Corporation for providing some nice pictures used in the book.

We are also indebted to the Dr. Rion Takahashi and the past and present students of our laboratory, who have contributed to this program.

Special thanks to Vassilios G. Agelidis, who encouraged the first author a few years ago to write a book on wind energy. Also acknowledged are Prof. Frede Blaabjerg, Dr. Stavros Papathanassiou, Dr. Vladislav Akhmatov, Dr. Zhe Chen, Dr. Gary L. Johnson, Dr. Mohammad Abdul Mannan, and Dr. Dharshana Muthumuni for providing different materials and supports that used in this book.

The authors are grateful to John Wiley & Sons, Taylor & Francis, Praise Worthy Prize, the Institution of Engineering and Technology, and the Institute of Electrical Engineers of Japan for permission to use materials from some of their renowned journals that were published earlier by the authors themselves.

The first author would like to thank Monbukagakusho (the Educational Ministry of Japan) for scholarship support during his Masters and PhD programs.

Finally, the authors acknowledge the Kitami Institute of Technology for supporting the research incorporated in this book.

Contents

Chapter 1

Introduction

1.1 Renewable Energy

Electricity is the most well-known energy carrier. An energy carrier is a substance or system that moves energy in a usable form from one place to another. Electricity is generated in power plants, in which a primary energy source is converted into electrical power. Examples of widely used primary energy sources are fossil fuels, falling or flowing water and nuclear fission. An important drawback of generating electricity from fossil fuels and nuclear fission, currently the most common primary energy sources for electricity generation worldwide, is the adverse environmental impact, such as the greenhouse effect caused by the increase of the CO_2 concentration in the earth's atmosphere and the nuclear waste problem. Further, fossil fuel and uranium reserves are finite. An additional disadvantage of using uranium and fossil fuels to generate electricity, particularly for those countries which do not have supplies of these primary energy sources, is the dependence on other countries for supplying a critically important resource.

In the 1970s, concern for the limited fossil fuel resources and their impact on the environment awakened. Due to this growing concern, interest revived in using renewable energy sources to meet the constantly rising world electricity demand. In addition, the oil crises of 1973 and 1979 led to the awareness that the amount of energy imported should be decreased so as to become less dependent on oil exporting countries. The Gulf-War (1990-1991) confirmed this concern.

The increasing concerns over environmental issues and the depletion of fossil fuel demanded the search for more sustainable electrical sources. One technology for generating electricity from renewable resources is to use wind turbines that convert the energy contained by the wind into electricity. The wind is a vast, worldwide renewable source of energy. Since ancient times, humans have harnessed the power of the wind. The earliest known use of wind power is the sailboat. Boats propelled by wind energy sailed up the Nile against the current as early as 5000 B.C. By A.D. 1000, the Vikings had explored and conquered the

North Atlantic. The wind was also the driving force behind the voyages of discovery of the Verenigde Oost-Indische Compagnie (VOC) between 1602 and 1799. Windmills have been providing useful mechanical power for at least the last thousand years, and wind turbines have generated electricity since 1888.

Compared to other renewable energy sources, such as photovoltaics, wave and tidal power, wind power is a relatively cheap source of renewable energy. Therefore, the promotion of renewable resources by a number of governments has led to the strong growth of wind power in the many countries.

1.2 Present Status and Future Prediction of Wind Power Worldwide

Globally, wind energy has become a mainstream energy source and an important player in the world's energy markets at the end of 2007. The following subsections will provide a condensed overview of wind energy status around the world until the year 2007. The prediction of future growth of wind energy until 2020 is also presented.

The Global Wind Energy Council (GWEC) has reported that 20,000 MW of wind power was installed in 2007, bringing world-wide installed capacity to 94,112 MW[1]. This is an increase of 31% over the 2006 market and represents an overall increase of about 27% in global installed capacity [1].

1.2.1 United States

The United States reported a record 5,244 MW installed in 2007, more than double the 2006 figure. This accounted for about 30% of the country's new power-producing capacity in 2007. Overall U.S. wind power generating capacity grew by 45% in 2007, with total installed capacity now standing at 16.8 GW. It can be expected that the U.S. will overtake Germany as the leader in wind energy by the end of 2009 [1].

The following map (Fig. 1.1) shows the installed megawatts (MW) for each state of the United States, as of mid January 2008 [2].

"Horse Hollow Wind Energy Center" is the world's largest wind farm at 735.5 MW capacity. It consists of 291 GE Energy 1.5 MW wind turbines and 130 Siemens 2.3 MW wind turbines spread over nearly 47,000 acres (190 km^2) of land in Taylor and Nolan Countries, Texas. The top ten ranking of the wind power generating states are shown in Table 1.1 [2]. Wind power generating capacity existing and under construction at the end of 2007 is shown in Table 1.2 [2].

---

[1] To obtain Global Wind Energy Council (GWEC) updated news, visit http://www.gwec.net/

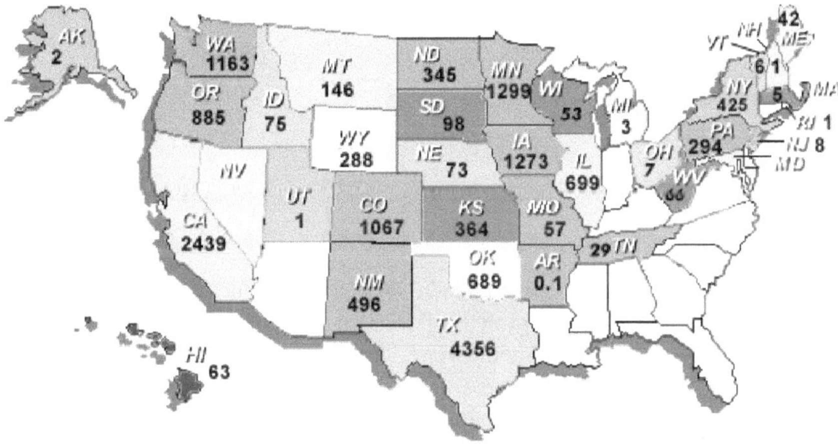

Fig. 1.1 AWEA Fourth quarter 2007 market report (Source: American Wind Energy Association, AWEA[2])

The U.S. Department of Energy has announced a goal of obtaining 6% of U.S. electricity from wind by 2020; a goal that is consistent with the current growth rate of wind energy nationwide [2].

Table 1.1 Top ten state total power capacity (MW)

State name	Existing	Under construction	Rank
Texas	4356.35	1238.28	1
California	2438.83	165	2
Minnesota	1299.75	46.4	3
Iowa	1273.08	116.7	4
Washington	1163.18	126.2	5
Colorado	1066.75	0	6
Oregon	885.39	15	7
Illinois	699.36	108.3	8
Oklahoma	689	0	9
New Mexico	495.98	0	10

Source: American Wind Energy Association, AWEA

[2] For the latest status of US wind energy installation, visit American Wind Energy Association (AWEA) at http://www.awea.org/

Table 1.2 National Total Power Capacities from Wind Energy (MW) in U.S.

Existing	Under Construction
16818.78	3506.38

Source: American Wind Energy Association, AWEA

1.2.2 Asia

GWEC [1] report says that China added 3,449 MW of wind energy capacity during 2007, representing market growth of 156% over 2006, and now ranks fifth in installed wind energy capacity with over 6,000 MW at the end of 2007. Based on current growth rates, the Chinese Renewable Energy Industry Association (CREIA) forecasts a capacity of around 50,000 MW by 2015.

The growing wind power market in China has also encouraged domestic production of wind turbines, and China now has more than 40 domestic companies involved in manufacturing. In 2007, domestic products accounted for 56% of the annual market, compared to 41% in 2006.

"This percentage is expected to increase substantially in the future. Total domestic manufacturing capacity is now about 5,000 MW and is expected to reach 10–12 GW by 2010," predicts GWEC President Prof. Arthouros Zervos.

India also continues to see a steady growth and now has about 8 GW of wind power installations, up from just over 6.2 GW in 2006.

Japan taking the third position in Asia, installed 1538 MW of wind energy capacity at the end of 2007.

1.2.3 Europe

The European Wind Energy Association (EWEA) published a report, which depicts a clear picture of wind energy installation in Europe through the end of 2007[3] [3]. The total capacity of new wind turbines brought on line across the European Union last year was 8,554 MW, an increase of 935 MW over the 2006 total. Total wind power capacity installed by the end of 2007 will avoid about 90 million tonnes of CO_2 annually and will produce 119 Terawatt hours in an average wind year, equal to 3.7% of EU power demand. In 2000, less than 0.9% of EU electricity demand was met by wind power.

"It is positive that wind energy is now increasing more than any other power technology in Europe. The market is up by 12% compared to 2006 but if we ex-

[3] For the updated status of European wind energy installation, visit the European Wind Energy Association (EWEA) at http://www.ewea.org/

clude Spain from the figures, the European market for wind turbines shows a small decline”, commented Christian Kjaer, EWEA Chief Executive.

Spain set a new record in 2007, installing 3,522 MW – the highest amount of any European country in any year ever. Now, 10% of its electricity comes from wind. There was also sustained growth in France – which added 888 MW to reach 2,454 MW – and in Italy, with 603 MW more and a total of 2,726 MW. The new Member States performed well and increased installed capacity by 60%, with Poland, the most successful, reaching a total of 276 MW. The Czech Republic installed 63 MW, its best year ever, and Bulgaria 34 MW.

Nevertheless, a handful of markets pulled in the opposite direction including Germany, Portugal, and the United Kingdom. As a result, the overall market growth in 2007 – 12% – was not as striking as it could have been. The global market for wind turbines grew by approximately 30% last year to 20,000 MW, and European companies continue to lead the market which is estimated to have been worth some €25 billion in 2007.

The change of pace in some countries can be explained by a mixture of slow administrative processes, problems with grid access and legislative uncertainty. “Spain – like Germany and Denmark before her – has taken the lead. There is no doubt in my mind that a swift approval by the 27 member states and the European Parliament of the Commission’s proposed renewable energy directive would pave the way for an equally massive expansion of wind energy in other member states”, said Christian Kjaer.

Wind power continued to be one of the most popular electricity generating technologies in the EU in 2007, making up 40% of total new power installations. Since 2000, the EU has installed 158,000 MW of new power capacity. New gas installations totalled 88,000 MW; wind energy 47,000 MW; coal 9,600 MW; oil 4,200; hydro 3,100; biomass 1,700 MW; and nuclear 1,200 MW during the eight-year period, according to figures from Platts PowerVision and EWEA.

Figure 1.2 shows the total wind power installation capacities throughout Europe at the end of 2007 [3].

1.2.4 Middle East and North Africa

Although Europe, North America and Asia continue having the largest additions to their wind energy capacity, the Middle East/North Africa region increased its wind power installations by 42%, reaching 534 MW at the end of 2007. New capacity was added in Egypt, Morocco, Iran, and Tunisia [1].

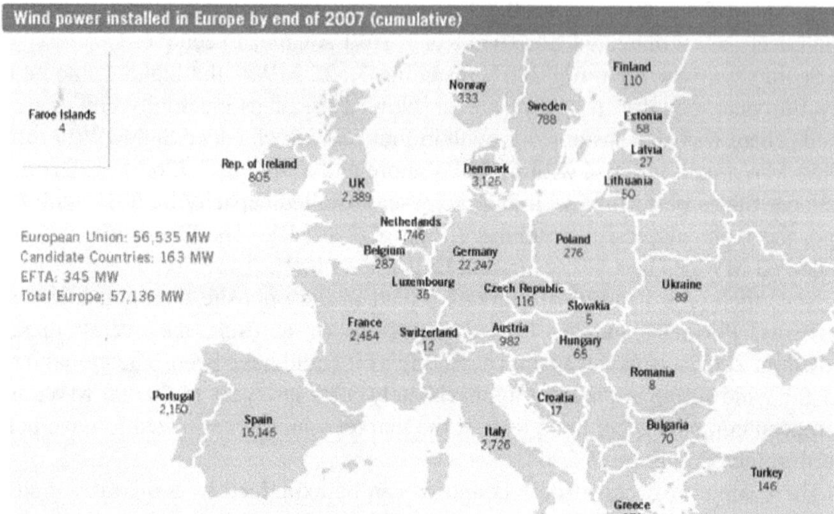

EU Capacity (MW)	End 2006	Installed 2007	End 2007
Austria	965	20	982
Belgium	194	93	287
Bulgaria	36	34	70
Cyprus	0	0	0
Czech Republic	54	63	116
Denmark	3,136	3	3,125
Estonia	32	26	58
Finland	86	24	110
France	1,567	888	2,454
Germany	20,622	1,667	22,247
Greece	746	125	871
Hungary	61	4	65
Ireland	746	59	805
Italy	2,123	603	2,726
Latvia	27	0	27
Lithuania	48	7	50
Luxembourg	35	0	35
Malta	0	0	0
Netherlands	1,558	210	1,746
Poland	153	123	276
Portugal	1,716	434	2,150
Romania	3	5	8
Slovakia	5	0	5
Slovenia	0	0	0
Spain	11,623	3,522	15,145
Sweden	571	217	788
UK	1,962	427	2,389
Total EU-12	**419**	**263**	**675**
Total EU-15	**47,651**	**8,291**	**55,860**
Total EU-27	**48,069**	**8,554**	**56,535**
Of which offshore	*870*	*210*	*1,080*

	End 2006	Installed 2007	End 2007
Candidate Countries (MW)			
Croatia	17	0	17
FYROM*	0	0	0
Turkey	50	97	146
Total	**67**	**97**	**163**
EFTA (MW)			
Iceland	0	0	0
Liechtenstein	0	0	0
Norway	325	8	333
Switzerland	12	0	12
Total	**337**	**8**	**345**
Other (MW)			
Faroe Islands	4	0	4
Ukraine	86	3	89
Total Europe	**48,563**	**8,662**	**57,136**

*FYROM = Former Yugoslav Republic of Macedonia

Note: Due to previous-year adjustments, project decommissioning of 88 MW, re-powering and rounding figures up and down, the total for the 2007 end-of-year cumulative capacity is not exactly equivalent to the sum of the 2006 end-of-year total plus the 2007 additions.

Source: EWEA

Fig. 1.2 Wind power installation statistics in Europe at the end of 2007 (Source: European Wind Energy Association, EWEA)

1.2.5 Pacific Region

After some slow years, the Pacific market gained new impetus in 2007, especially in New Zealand, where 151 MW were installed in 2007. In Australia, the newly elected Labour government has ratified the Kyoto Protocol and pledged to introduce a 20% target for renewable energy by 2020, justifying an optimistic outlook for future wind energy developments [1].

1.2.6 Latin America and Caribbean Region

The Latin America and the Caribbean region had the slowest growth rate compared to the rapid growth of wind power generation around the world. In Latin Americ, only Brazil installed 10 MW new wind power-generating stations in 2007, which was the highest in that region.

The top five countries in terms of installed capacity are Germany (22.3 GW), the United States (16.8 GW), Spain (15.1 GW), India (8 GW), and China (6.1 GW). In terms of economic value, the global wind market in 2007 was worth about 25 billion EUR or 36 billion US$ in new generating equipment [1].

The wind power installation throughout the world at the end of 2007 is shown in Fig. 1.3 [1]. Figure 1.4 gives a clear view of the statistics of world wind power installations [1].

1.3 Wind Turbine Technical Overview

EWEA has reported the wind turbine development history from 1980 to 2007 on its twenty-fifth anniversary [4]. The report is presented briefly in this section.

There have been attempts to combine more than one turbine on the same support system. In the 1980s, for example, the Dutch Lagerwey company installed a prototype of the 300 kW Quadro-four 75 kW two-bladed turbines, attached to the same tower. It weighed 400 tonnes. According to technical writer Eize de Vries, "after some teething problems the installation performed well for about 15 years". Other similar multi-rotor systems have been proposed but few were built.

Eventually, by the mid-1990s, the stage was set for the emergence of the most popular current design-a three bladed turbine with pitch (rather than stall) blade control, variable speed and a gearbox. Only Enercon of the largest manufacturers uses direct drive (gearless) operation.

Wind turbines have also steadily increased in both the extent of their rotor blade sweep and in their installed capacity-a gradual learning process rather

GLOBAL INSTALLED WIND POWER CAPACITY (MW) – REGIONAL DISTRIBUTION

		Revised end 2006	New 2007	Total end 2007
AFRICA & MIDDLE EAST	Egypt	230	80	310
	Morocco	54	60	114
	Iran	47	19	66
	Tunisia	20	0	20
	Other [1]	18	0	18
	Total	**369**	**159**	**528**
ASIA	India	6,270	1,730	8,000
	China	2,604	3,449	6,050
	Japan	1,394	139	1,538
	Taiwan	188	100	282
	South Korea	173	18	191
	Philippines	25	0	25
	Other [2]	5	0	5
	Total	**10,659**	**5,436**	**16,091**
EUROPE	Germany	20,622	1,667	22,247
	Spain	11,623	3,522	15,145
	Denmark	3,136	3	3,125
	Italy	2,123	603	2,726
	France	1,567	888	2,454
	UK	1,962	427	2,389
	Portugal	1,716	434	2,150
	Netherlands	1,558	210	1,746
	Austria	965	20	982
	Greece	746	125	871
	Ireland	746	59	805
	Sweden	571	217	788
	Norway	325	8	333
	Belgium	194	93	287
	Poland	153	123	276
	Rest of Europe [3]	556	263	812
	Total Europe	**48,563**	**8,662**	**57,136**
	of which EU-27 [4]	*48,069*	*8,554*	*56,535*
LATIN AMERICA & CARIBBEAN	Brazil	237	10	247
	Mexico	87	0	87
	Costa Rica	74	0	74
	Caribbean (w/o Jamaica)	35	0	35
	Argentina	20	0	20
	Colombia	27	0	27
	Jamaica	20	0	20
	Chile	2	18	20
	Cuba	5	0	5
	Total	**507**	**28**	**535**
NORTH AMERICA	USA	11,575	5,244	16,818
	Canada	1,460	386	1,846
	Total	**13,035**	**5,630**	**18,664**
PACIFIC REGION	Australia	817	7	824
	New Zealand	171	151	322
	Pacific Islands	12	0	12
	Total	**1,000**	**158**	**1,158**
	World total	**74,133**	**20,073**	**94,112**

1 Cape Verde, Israel, Jordan, Nigeria, South Africa;
2 Bangladesh, Indonesia, Sri Lanka;
3 Bulgaria, Croatia, Cyprus, Czech Republic, Estonia, Faroe Islands, Finland, Hungary, Iceland, Latvia, Liechtenstein, Lithuania, Luxembourg, Malta, Norway, Romania, Russia, Slovakia, Slovenia, Switzerland, Ukraine;
4 Austria, Belgium, Bulgaria, Cyprus, Czech Republic, Denmark, Estonia, Finland, France, Germany, Hungary, Ireland, Latvia, Lithuania, Luxembourg, Malta, Netherlands, Poland, Portugal, Romania, Slovakia, Slovenia, Spain, Sweden, UK;

Please note: project decommissioning of 93MW and rounding affect the final sums

Source: GWEC

Fig. 1.3 Wind power installation statistics throughout the world at the end of 2007 (Source: Global Wind Energy Council, GWEC)

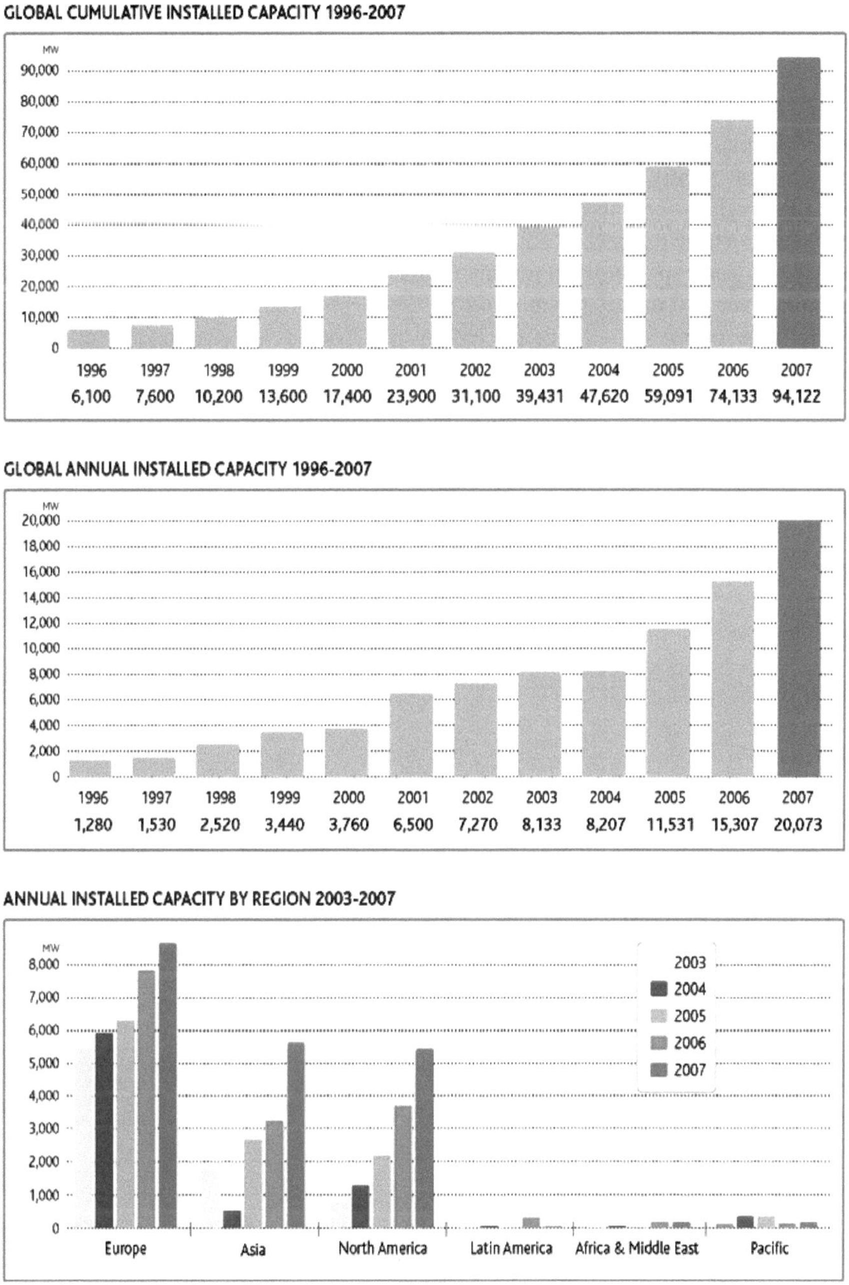

Fig. 1.4 Wind power installation statistics throughout the world at the end of 2007 (Source: Global Wind Energy Council, GWEC)

than the giant leaps of the 1980s. By 1996, for example, Wind Directions reported that most of the leading Danish manufacturers, including Nordex, Vestas, Nordtank and Bonus, were producing 1 to 1.5 MW capacity turbines. The 1.5 MW machines have continued to be the most popular rating well into the current century. By 2005, more than 3,000 of the 1.5 MW model originally designed by the German family company Tacke Windtechnik, and eventually taken over by GE Energy, had been installed, whilst Enercon had produced 2,400 of its E-66 1.5 – 2 MW capacity turbines.

The largest capacity wind turbine on the market now is 5 MW, the tallest towers well over 100 meters (mainly to catch the wind at inland sites) and the longest blades over 60 meters – twice the length of a Boeing 747 jumbo jet. Turbines of this size are designed primarily for use in the growing offshore market, where the cost of foundations and electrical connection makes it sensible to use fewer more powerful machines. Figure 1.5 shows the development of the wind turbine size between 1980 – 2003 [4].

ENERCON is currently installing two E-126/6 MW WECs on the Rysumer Nacken in Emden, Germany[4]*. This new ENERCON model is a sophisticated version of the E-112 (6 MW rated power) – the world's most powerful wind turbine to date [5].*

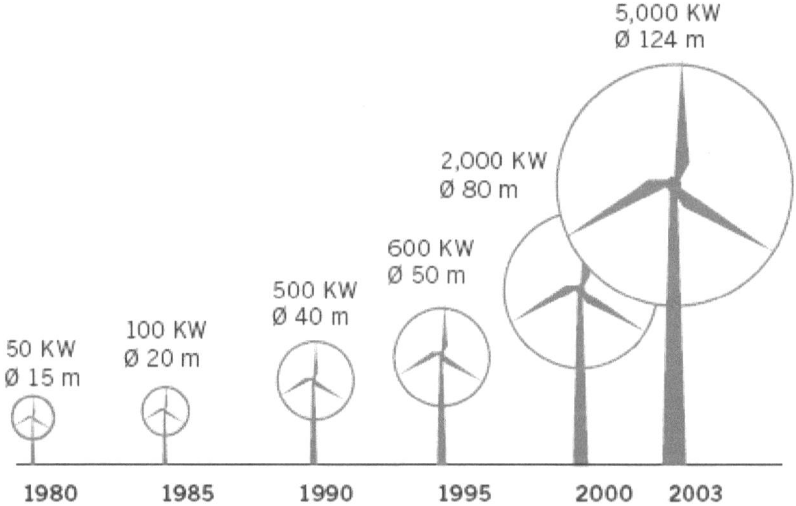

Fig. 1.5 Development of wind turbine size, Source: Wind Energy-The Facts (European Wind Energy Association, EWEA, 2004)

[4] For the latest information about ENERCON, visit at http://www.enercon.de/en/_home.htm

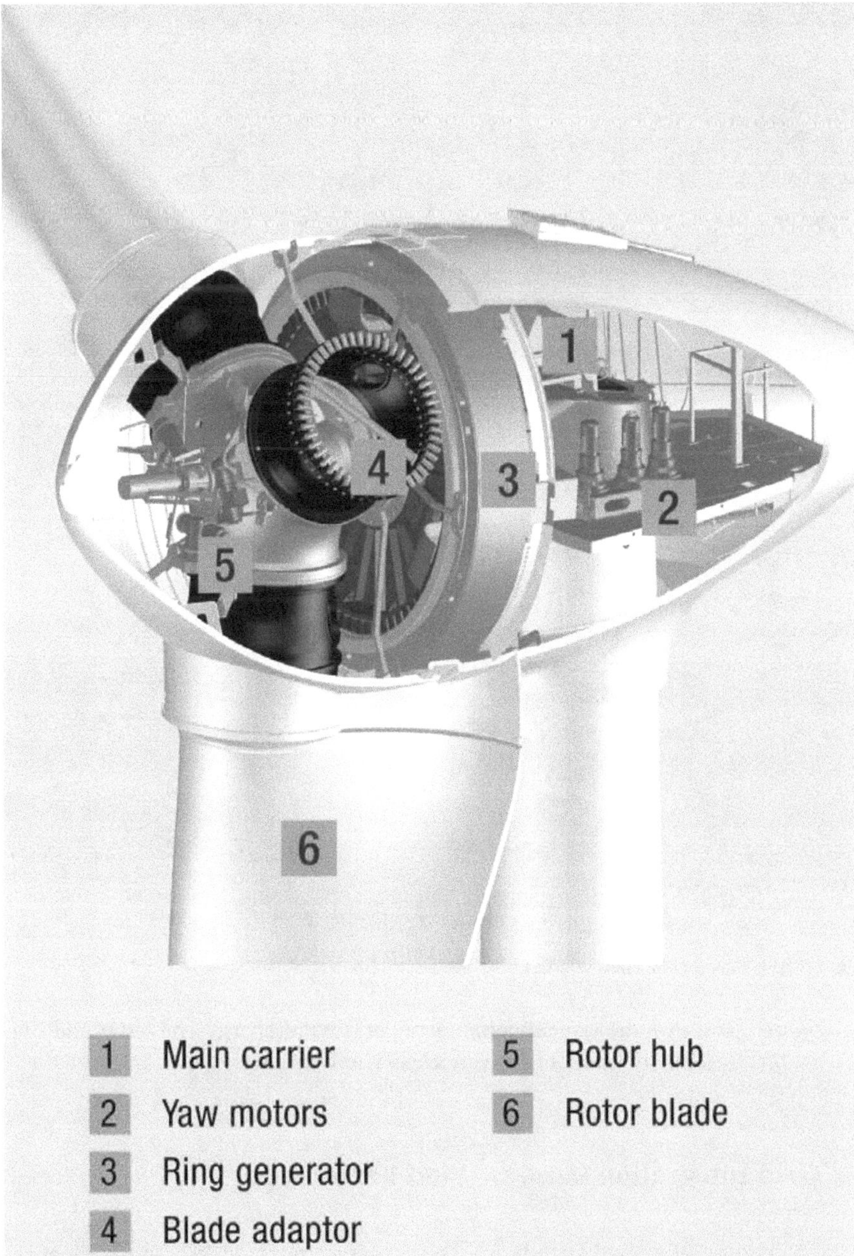

Fig. 1.6 Enercon E-82 wind energy converter (Source: ENERCON GmbH)

Fig. 1.7 REpower 5 MW class wind turbine (Source: REpower Systems AG)

Figure 1.6 shows the typical components of a wind energy converter and Fig. 1.7 is a photography of a modern 5-MW class wind turbine.

1.4 Grid Integration Issue of Wind Farm

Grid integration of a wind farm is an important issue now a days. An excellent report on this topic published in 2005 by EWEA is presented below [6].

Often grid codes contain very costly and challenging requirements that have no technical justification. They are often developed by vertically-integrated power companies, i.e., within companies in competition with wind farm operators, in highly non – transparent manners. Furthermore, there are continuous changes in grid codes, technical requirements and related regulation, often introduced on very short notice and with minimum involvement of the wind power sector.

The general frameworks for integrating wind power should acknowledge that technical requirements – such as grid codes, curtailment practices, reactive power etc. – depend to a large extent on the wind power penetration levels and the nature of the existing infrastructure, e.g., interconnectors and the overall generation mix.

Grid codes and other technical requirements should reflect the true technical needs and should be developed in cooperation with independent and unbiased transmission systems owners (TSO), the wind energy sector, and independent regulators.

Grid codes and grid access requirements should take into account that, at low penetration levels, excessive requirements such as fault ride through capability and voltage control possibilities are often imposed on wind power generators without being technically justified. Costly requirements should be included only if they are technically required for reliable and stable power system operation. The assessment of the requirements should be made by independent bodies – not by transmission operators that are affiliated with vertically integrated power producers.

A large geographical spread of wind power on a system should be encouraged through planning and payment mechanisms and the establishment of adequate interconnection. From a system and cost point of view that will reduce variability, increase predictability and decrease or remove situations of near zero or peak output. The cost of grid extension should be socialized. One reason to do it is that grids are natural monopolies. Grid connection charges should be fair and transparent and competition should be encouraged. In future developments of the European power systems, increased flexibility should be encouraged as a major design principle (flexible generation, demand side management, interconnections, storage etc.). In addition, public private partnership and use of structural funds should play an important part. The benefits of distributed generation, e.g., reduced network losses and reduced need for grid reinforcements, must be recognized [6].

1.5 Background of the Book

Two types of wind turbine generator systems (fixed and variable speed) are commercially available. Due to the current interest in wind energy generation, it is predicted easily that huge numbers of wind farms are going to be connected to the existing network in the near future. Therefore, it is necessary to analyze both the steady state and transient stability of grid connected wind farms.

In general, all generators connected to the transmission system are required to comply with a grid code. The grid codes were originally decided with synchronous generators in mind. But due to the recent addition of huge amounts of wind power to the grid, new grid codes have been developed in many countries to ensure secure power system operation [7 – 9]. For example, in Germany, the wind generator shut down phenomenon has been reduced by adopting the low voltage ride through (LVRT) requirement from the German grid operator named E.ON Netz. The E.ON Netz standard requires that the machine remains connected to the grid if the terminal voltage is higher than 0.15 pu for approximately 0.6 sec [7]. American Wind Energy Association (AWEA) is also recommending adopting the LVRT requirement developed by E.ON Netz. In [8, 9], it is stated that wind farm terminal voltage has to return to 90 % of the nominal voltage within 3 seconds after the starting of a voltage drop. Otherwise, the plant has to be shutdown. However, wind farm compatible grid codes are more or less similar. The fact is that wind farm has to comply with the grid code to be connected to the existing power grid.

Induction generators are used widely, in general, as fixed speed wind generators due to their superior characteristics, such as brushless and rugged construction, low cost, maintenance free, and operational simplicity. But they have some stability problem as reported in [10]. The squirrel cage induction generator requires large reactive power to recover the air gap flux when a short circuit fault occurs in the power system. If sufficient reactive power is not supplied, then the electromagnetic torque of the induction generator decreases significantly. Then the difference between mechanical and electromagnetic torques becomes large, and the induction generator and turbine speeds increase rapidly. As a result, the induction generator becomes unstable and it requires to be disconnected from the power system. To enhance the fault ride through capability of fixed speed wind generator, different types of tools are used with wind turbine generator system (WTGS).

Before discussing transient stability or fault ride through capability enhancement of fixed speed WTGS, the drive train modeling of a WTGS should be considered. There are several reports investigating the transient characteristics of the fixed speed WTGS in faulted conditions [11 – 14]. In these references, however, the wind turbine and wind generator are modeled as a one-mass lumped model having a combined inertia constant. Because the one-mass lumped model is too simple to represent the dynamics of a wind turbine and wind generator connected to each other through a shaft with low stiffness, stability analysis based on the one-mass shaft model may result in significant error. The WTGS is the only generator unit in the utility network where mechanical stiffness is lower than electrical stiffness (synchronizing torque coefficient) [15]. Moreover, inertia constants of the turbine and generator have significant effect on the transient stability. Valuable studies have been performed for transient stability, fault analysis, and reactive power compensation of WTGS using a two-mass shaft model [15 – 26]. In other studies [27 – 32], three-mass or higher order drive train models are also analyzed. Flicker, power fluctuation, and torsional natural frequency of wind generators are discussed in [21, 30], where multi-mass drive train model is also considered. But from these reports, the characteristics of accuracy of each WTGS model, i.e., six-mass, three-mass, and two-mass drive train models, cannot be determined

The next issue is to stabilize a WTGS by using a pitch controller. The main purpose of the pitch controller is to maintain the output power of a wind generator at its rated level when the wind speed is above the rated speed. Besides this, it can enhance the transient stability of a WTGS by controlling the rotor speed. In some previous works [13, 14, 33, 34], it has been proposed that the pitch controller control the rotor speed when severe network disturbances occur in the power system. In those works, rotor speed was chosen as the pitch controller input. The control methodology of the pitch controller by taking speed and power change as inputs is discussed in [35], though the power and speed control modes are not distinguished there. Other valuable studies have been carried out in [20, 36 – 42] where the pitch controller is analyzed for a WTGS. Some of them also contain technical details of the pitch controller mechanics. But none showed the power and speed control modes of the pitch controller together. In [43], though the power and speed control modes are considered, but the controller is not robust for all types of operating conditions.

Reactive power compensation is a major issue, especially for a fixed speed WTGS in both steady state and transient conditions. Usually, a capacitor bank is placed at each induction generator terminal to provide the necessary reactive power in the steady state. But it cannot maintain the constant wind generator terminal voltage under randomly fluctuating wind speed. Moreover, the induction generator needs large reactive power during short circuit faults, as explained before. Recently voltage-source or current-source inverter based flexible AC transmission systems (FACTS) devices such as static var compensator (SVC), static synchronous compensator (STATCOM), dynamic voltage restorer (DVR), solid state transfer switch (SSTS) and unified power flow controller (UPFC) have been used for flexible power flow control, secure loading and damping of power system oscillation [44 – 46]. Some of those can be used to improve the transient and dynamic stability of wind generator. SVC is reported with a synchronous generator in [47] and with induction generator in [48] for reactive power compensation. But STATCOM has somewhat better performance compared to SVC for reactive power compensation, which is reported clearly in [49, 50]. Modeling of voltage source converter (VSC) based STATCOM is discussed in [51 – 63]. In [51 – 53, 55 – 60], a two level VSC based STATCOM is discussed. But for high voltage application, at least a three-level inverter is the right choice for a VSC based STATCOM [61 – 63]. Some authors have reported valuable studies on a STATCOM connected WTGS [56 – 60]. In [56], steady state reactive power control and islanding performance of induction generator are discussed. In [57], it is reported that a STATCOM can recover the terminal voltage of a wound rotor induction generator after the fault clearance. But as only an induction generator is connected to the network, the effect of the STATCOM on the rest of the system is not presented there. Flicker mitigation of a wind generator by using the STATCOM is discussed in [58 – 59]. In [60], a one-mass lumped model of the WTGS is considered instead of a two-mass drive train model. Though many works with a STATCOM are reported in power system literature, the stability analysis of the WTGS by using a STATCOM is not sufficient. Another point is that the damping characteristics of the shaft torsional oscillation of a steam turbine generator system are presented in many papers [64, 65]. But the damping of the blade-shaft torsional oscillations of a fixed speed WTGS during short circuit faults has so far not been reported in literature. A STATCOM can be an effective tool for minimizing the blade-shaft torsional oscillations of fixed speed WTGS.

Though wind power is considered a very prospective energy source, wind power fluctuation due to randomly varying wind speed is still a serious problem for power grid companies or transmission system owners (TSO), especially for fixed speed wind generators. The wind power fluctuation usually occurs on the timescale of a few seconds to several hours, depending on the wind condition, wind turbine size, topology, etc. The wind power fluctuations are comparatively smaller in a wind farm than in a single WTGS on these timescales. But considering the isolated systems or as an option for future energy systems with high penetration of wind power generation, it is essential to emphasize research on wind power smoothing. This book focuses on wind power fluctuation on the timescale of minutes. In [66 – 68], a flywheel energy system is proposed for smoothing the wind power fluctuations. A flywheel system has high standby loss within the range of 5% of its power rating. Moreover, the control strategy is complicated. It is also possible to smooth wind power fluctuations up to a certain range by pitch angle control [69 – 72]. However, the reported pitch controller needs complex computation resulting in an increase in the cost of the controller. Some authors have proposed [73 – 75] a superconducting magnetic energy storage (SMES) system for minimizing wind power fluctuation. Though SMES is a very good system for wind power smoothing due to its high response speed and high efficiency, its practical implementation in large megawatt range applications is still doubtful for its large installation and continuous maintenance cost. In some reports, a battery energy storage system (BESS) integrated with a STATCOM is proposed to obtain real and reactive power support [76 – 79]. A STATCOM/BESS can also be used for wind power smoothing [80]. But the application of STATCOM/BESS in wind power application may not be a good choice because it is based on a chemical process and it has low response speed and a short service life. A nice study has already been reported, where the merits and limitations of different types of energy storage devices suitable for wind power application are discussed in detail [81].

Another recently developed technology is an energy capacitor system (ECS), which combines power electronic devices and an electric double layer capacitor (EDLC). It has both real and reactive power controllability [82 – 90]. This system features "clean energy" from an environmental point of view. The EDLC has a simple charging method; there is no need to build any protective circuits. After a full charge, it stops accepting charge. An EDLC can be cycled millions of times, i.e., it has a virtually unlimited cycle life. Its standby loss is very low within the range of 0.2% of its power rating. Therefore, an ECS can be used effectively for wind farm power smoothing. A nice study is presented in [90], where ECS is used for wind power smoothing, though real wind speed data are not used there. Moreover, in that study, there is no indication of how to determine the reference power for the control scheme used in an ECS. Wind is intermittent and stochastic by nature. Therefore, an appropriate control strategy must be adopted along with suitable reference line power to minimize the output power fluctuation from a wind farm. Additionally, an ECS can be used to enhance the low voltage ride through (LVRT) capability of a wind farm because it has reactive power controllability.

Hydrogen has received much attention recently because of the limited supplies of fossil fuel and global warming. One recent trend is to generate hydrogen by using wind energy. Wind power generation is not possible all the time and there are places where sufficient wind speed is not available. But if it is possible to transform wind energy into hydrogen and to store the hydrogen, then it can be kept in a stable state for a long time. In that case, the hydrogen can also be transported easily to any places. In [91], the hydrogen generation topologies suitable for variable speed wind generators were discussed. Detailed hydrogen storage methodology was discussed in [92]. In [93 – 94], hydrogen generation from wind energy was presented for stand-alone system. But the study of constant hydrogen generation from grid connected fixed speed wind farm is not sufficient enough.

The variable speed wind turbine (VSWT) has recently become more popular than the fixed speed WTGS. Doubly fed induction generators (DFIG), wound field synchronous generators (WFSG), and permanent magnet synchronous generators (PMSG) are currently used as variable speed wind generators. The stability analysis of the DFIG type of the VSWT is already reported in many literatures [95 – 97]. On the other hand, though the stability analysis of synchronous generators has been done extensively in many papers [98, 99], it is quite insufficient when the synchronous generator is used for wind power generation with the full rating of power converter topology. In [100], the transient stability analysis of a VSWT using a field excited synchronous generator has been presented, where only an unsymmetrical fault is considered as a network disturbance. On the other hand, the transient stability analysis of a VSWT driving a PMSG is not sufficient. In a PMSG, the excitation is provided by permanent magnets instead of a field winding. Permanent magnet machines are characterized as having large air gaps, which reduce flux linkage even in machines with multi-magnetic poles [101 – 102]. As a result, low rotational speed generators can be manufactured with relatively small sizes with respect to their power rating. Moreover, gearboxes can be omitted due to low rotational speed in PMSG wind generators, resulting in low cost. In some survey studies, gearbox is found to be the most critical component, since its downtime per failure is high in comparison to other components in WTGS. In [103 – 105], detailed modeling and control systems for variable speed PMSG wind generator systems are presented, but in those papers the steady state and dynamic performances are not analyzed sufficiently. In [106], the transient characteristic of a VSWT-PMSG is discussed for only a step change of generator speed, but no fault condition is considered there. More research should be performed in controller design and stability analysis of the VSWT-PMSG.

1.6 Scope and Aims

In view of what is presented above in the area of wind farm stability enhancement, several vital studies still need to be investigated.

1) This book compares the results among different types of drive train models of the fixed speed WTGS, which will be very effective for transient stability analysis. It will be shown that the reduced order two-mass shaft model is sufficient enough for the transient stability analysis of a fixed speed WTGS. In order to distinguish the characteristics between the six-mass model and reduced order models, i.e., three and two mass drive train models, proper definitions of all concepts under comparison must be made first. Next, all concepts under consideration should have parameters that can be obtained using the proper transformation technique from a base parameter set of the six-mass drive train model, which is considered a bench mark model. Finally, all concepts under consideration should be compared to each other under exactly the same operating conditions. Some reports on this issue are already published in [107, 110, 111].

2) Later, in this book, a new logical pitch controller has been proposed that works in both power and speed control modes. To obtain robust performance from the pitch controller it is necessary to consider the terminal voltage of a wind generator as one of the pitch controller inputs. Depending on the network parameters or conditions, there might be a situation where the pitch controller with rated reference power cannot stabilize the wind generator. And it is usual that at lower terminal voltage the wind generator cannot generate its rated power. Therefore, it is logical to change the reference power of the pitch controller according to the terminal voltage of the wind generator. Then the system would be more stable and robust for any operating condition. With this view, a new logical pitch controller has been proposed in this work. The proposed controller has a simple logical unit that can determine the operating status of the controller in a very effective and logical way. Another feature of this pitch controller is that the mechanical dead zone of the pitch actuation system is also considered in the pitch controller design phase to obtain a realistic response. All simulation results related to the pitch controller include the dead time, which makes the proposed controller applicable to a real system. As a control methodology of the proposed pitch controller, the fuzzy logic controller (FLC) is used. The performance of the FLC is compared to that of the conventional PI (proportional-integral) controller. Both type of pitch controllers are reported in our earlier work [108].

3) In the previous works with the STATCOM and WTGS [55 – 60], multi-mass shaft model was not considered. But in [22, 23, 26], it is reported that for transient stability analysis of a fixed speed WTGS, a multi-mass shaft model should be considered, though no types of reactive power compensation tools were used in those studies. The reactive power compensation of a wind generator with a static shunt capacitor is reported in [24], where a two-mass shaft model was considered. But in that paper, fault-clearing time was emphasized. Therefore, in this

book, we propose a three level STATCOM based on a voltage source converter PWM technique to stabilize a fixed speed WTGS, considering the two-mass drive train model. Detailed modeling and control strategy for a VSC-based STATCOM are presented. Both symmetrical and unsymmetrical faults are analyzed. Moreover, it is presented that a STATCOM can reduce wind generator voltage fluctuation under variable wind speed, i.e., STATCOM can improve the power quality of a fixed speed WTGS. Finally, it is shown that besides a WTGS, a STATCOM can also enhance the stability of an entire power system including the synchronous generator. This topic is presented in the authors' earlier work [109].

4) In addition to the transient stability enhancement of a WTGS by using STATCOM, in this book, the minimization of the blade and shaft torsional oscillations of a fixed speed WTGS are also described by using both the pitch controller and the STATCOM. Here, the six-mass drive train model is considered for the sake of precise analyses. It is shown how much blade-shaft torsional oscillations can be minimized by using mechanical input torque control and electromagnetic torque restoration during network disturbances from using the pitch controller and the STATCOM respectively. Damping of the blade-shaft torsional oscillation of a fixed speed WTGS is reported in [112].

5) It is reported that pitch controller can also be used to smooth the wind generator output power. The fuzzy logic controlled pitch controller is proposed to smooth the wind generator output power. Three different types of averages are considered to determine the controller input reference command power, and finally, the exponential moving average (EMA) is chosen as the reference of the wind generator line power. Power smoothing by using the pitch controller is presented in our previous work [113].

6) In this book, it is reported that ECS topology can significantly decrease the power fluctuation of a grid connected fixed speed wind generator system. The proposed ECS topology can also be applied with a variable speed wind generator. One major problem in smoothing a wind generator output power is the determination of the line power reference of a wind farm. In some countries, the TSO or grid authority provides the wind farm grid code. The wind farm grid code related to the voltage level reference or the low voltage ride through capability can be set easily. But the reference power of the wind farm may not be set suitably as wind is a randomly varying entity. Moreover, the reference power of each individual wind turbine may increase the complexity of the control system. Therefore, we focused the line power smoothing of a wind farm, based on the reference line power. However, constant line power reference is not a good choice. In that case, the capacity of the ECS system must be extremely large. This paper proposes using the exponential moving average (EMA) to generate the line power reference. By adopting this method, the capacity of the ECS unit can be decreased. The objective of the control system is to follow the reference line power by absorbing or providing real power. The control scheme of an ECS is based on a sinusoidal PWM (pulse width modulation) voltage source converter (VSC) and DC-DC buck/boost converter using the insulated gate bipolar transistor (IGBT).

On the other hand, another major problem in wind power generation from a fixed speed WTGS is the terminal voltage fluctuation. Generally, a capacitor bank system is installed at the terminal of a fixed speed wind generator, which cannot maintain its terminal voltage at the desired reference level when the wind speed fluctuates randomly. Usually, the wind farm terminal voltage is not kept constant at the rated voltage, but is reset to a desired value once or a few times a day. Since the proposed ECS system can also provide necessary reactive power to wind generators, the wind farm terminal voltage can be kept at a desired reference level. Simulation results clearly show that our proposed control system can smooth well the wind farm line power, keeping its terminal voltage at the desired reference level under randomly fluctuating wind speed. This topic is partly reported in our previous work [114].

7) In recent years, hydrogen generation from wind power has become popular. A hybrid control system composed of a fixed speed WTGS, an ECS, and a hydrogen generator is presented and a cooperative control system is designed. An economical hydrogen generator topology with an ECS connected at the wind farm terminal is proposed.

8) In addition, the low voltage ride through capability enhancement of a fixed speed WTGS by using an ECS is presented. It is reported that an ECS can significantly augments the transient stability of a power system including wind farms. This topic is also partly reported in the authors' earlier work [115].

9) In this book, the transient stability of a VSWT driving a PMSG is analyzed in detail. For the maximum power point tracking (MPPT) operation, rotor speed is used as a controller input instead of wind speed, because the rotor speed can be measured precisely and more easily than the wind speed. In the model system, a PMSG wind generator is connected to the power system network through a fully controlled frequency converter. Two types of electrical schemes for the frequency converter are considered. One consists of generator side AC/DC converter, DC link capacitor, and grid side DC/AC inverter. The second one consists of rectifier, boost converter, DC link capacitor, and grid side DC/AC inverter. The detailed control schemes of both frequency converters are presented. Simulation analyses have been performed by using PSCAD/EMTDC, in which the PMSG, electrical network, both converter/inverter, and their controllers are implemented by using standard library models. Different types of symmetrical and unsymmetrical faults are considered to be occurred at several locations in the model system. It is found that by controlling the power converters of the PMSG properly, the transient stability of the VSWT-PMSG can be enhanced.

This book presents contributions to solving the problems given above (1 – 9). For the sake of precise analysis real wind speed data measured on Hokkaido Island, Japan are used when needed. For transient stability analyses both balanced (3LG: three-line-to-ground) and unbalanced (1LG: single-line-to-ground, 2LG: double-line-to-ground, and 2LS: line-to-line) faults are considered to occur in the power system including a wind farm. Some portions of those issues have already been reported in some papers [107 – 115].

The simulations have been carried out with the popular power system simulator software package PSCAD/EMTDC. In some cases, a FORTRAN program is also incorporated with PSCAD/EMTDC to implement some new modeling and control strategy, not available in the standard library.

1.7 Outline of this Book

Chapter 2 provides a general overview of the commercially available wind turbine topologies. Both fixed and variable speed wind turbine characteristics are presented, which are used in the rest of this book. The six-mass, three-mass, and two-mass drive train models of a fixed speed WTGS are presented in detail. The transformation methodology from the six-mass to two-mass drive train models is presented. This can be used in the simulation analysis with reasonable accuracy. The effects of drive train parameters such as inertia constants, spring constants, and damping constants are examined for the above mentioned three-types of drive train models. Fixed and variable speed WTGS characteristics, drive train modeling, and topological overview are presented briefly in this chapter.

In Chap. 3, two types of pitch controllers are reported. First, a new logical pitch controller equipped with a fuzzy logic controller (FLC) is presented. This FLC controlled pitch controller can maintain the output power of a wind generator at rated level when the wind speed is above the rated speed. In addition, it can enhance the transient stability of a WTGS during severe network disturbances. To obtain robust performance, the wind generator terminal voltage is taken as the pitch controller input. The performance of a FLC is compared with that of a PI controller. Another new feature of this chapter is the wind generator power smoothing by using a pitch controller. Three different types of average values, i.e., average (AVG), simple moving average (SMA), and exponential moving average (EMA) are analyzed to generate the pitch controller input power command, and finally, the EMA is recommended. The FLC is proposed as the control methodology of the pitch controller for wind power smoothing. Some mechanical aspects such as mechanical dead zone, rate limiter, etc., are also considered for both of the pitch controllers, which make the controllers practically applicable.

Chapter 4 presents the detailed modeling and control strategy for a three level voltage source converter based STATCOM to enhance both the steady state and transient performances of grid connected fixed speed WTGS. For the transient performance evaluation, a two-mass drive train model of WTGS is considered, as reported in Chap. 2. It is reported that a STATCOM can improve the power quality of a grid connected wind generator in steady state operation. Moreover, the minimization of the blade-shaft torsional oscillations of a fixed speed WTGS during network disturbances is also reported.

Chapter 5 emphasizes the energy storage systems (ESSs). First, a general discussion on ESSs for power system application is presented. Then four types of

ESSs suitable for wind power application are described in detail. The topological overviews of STATCOM incorporated with battery energy storage system (STATCOM/BESS), flywheel energy storage system (FESS), superconducting magnetic energy storage system (SMES), and energy capacitor system (ECS) for wind power application are presented. Finally, a cost comparison is made in the light of the report by Sandia National Laboratories [132].

In Chap. 6, first a general discussion on hydrogen gas is presented including its application in a different sector. Then hydrogen generation by using wind power is emphasized. The basic modeling of a hydrogen generator including the electrolyzer function is presented. The hydrogen storage procedure is incorporated in the light of the report from National Renewable Energy Laboratory (NREL) [138].

In Chap. 7, a new wind farm operating strategy is presented using a wind farm, an ECS, and a hydrogen generator. The detailed modeling and control strategy for an ECS including its individual components are presented. It is reported that an ECS can smooth the line power and terminal voltage of a fixed speed wind farm, because it has both real and reactive power controllability. Additionally, by taking advantage of an ECS, the economical and performance effective hydrogen generator topology integrated at the wind farm terminal is proposed. It is also reported that an ECS can enhance the low voltage ride through capability of a fixed speed wind farm. Moreover, it is reported that an ECS can enhance the transient stability of a power system including wind farms.

Chapter 8 presents a comprehensive study of the transient stability of a variable speed wind turbine driving a PMSG when a network disturbance occurs in the power system. Detailed modeling and a control strategy for two types of electrical schemes of the frequency converter are presented. The proposed control strategies can provide maximum power to the grid and can also control the reactive power to maintain the terminal voltage of the grid constant. It is shown that the control strategies of both frequency converter topologies can give excellent transient performance under both symmetrical and unsymmetrical fault conditions even at different locations in the power system.

Chapter 2

Wind Turbine Modeling

This chapter describes the details of wind turbine modeling. First, a wind energy conversion system is discussed briefly. Then, the modeling and topological overview are presented for both fixed and variable speed wind turbine generator systems. The details of drive train modeling of a fixed speed WTGS are discussed and a six-mass to two-mass conversion method is also presented. Finally, a comparative study is carried out among six-mass, three-mass, and two-mass drive train models.

2.1 Wind Power Output

Gathering or harvesting the wind has been of concern to humans for a long time. Wind turbines have been used for several centuries and literally millions of units have been put into service. For the most part, these machines performed their intended purpose well and in many cases were still being used with minimum maintenance after a half century of service.

Today, wind turbines have to compete with many other energy sources. Therefore, it is important that they should be cost effective. They need to meet any load requirements and produce energy at a minimum cost per dollar of investment. Performance characteristics such as power output versus wind speed or versus rotor angular velocity must be optimized to compete with other energy sources. Yearly energy production and its variation with annual wind statistics must be well known. The shaft torque must be known so that the shaft can be built with adequate strength and the turbine load is properly sized. The output power from ideal and practical wind turbines is discussed in the following two sections [116].

2.1.1 Power Output from an Ideal Turbine

The kinetic energy in a parcel of air of mass, m, flowing at speed, v_w in the x direction is:

$$U = \frac{1}{2} m v_w^2 = \frac{1}{2} (\rho A x) v_w^2 \tag{2.1}$$

where, U is the kinetic energy in joule, A is the cross-sectional area in m^2, ρ is the air density in kg/m^3, and x is the thickness of the parcel in m. If we visualize the parcel as in Fig. 2.1 with side, x, moving at speed, v_w (m/sec), and the opposite side fixed at the origin, we see the kinetic energy increasing uniformly with x, because the mass is increasing uniformly.

The power in the wind, P_w, is the time derivative of the kinetic energy:

$$P_w = \frac{dU}{dt} = \frac{1}{2} \rho A v_w^2 \frac{dx}{dt} = \frac{1}{2} \rho A v_w^3 \tag{2.2}$$

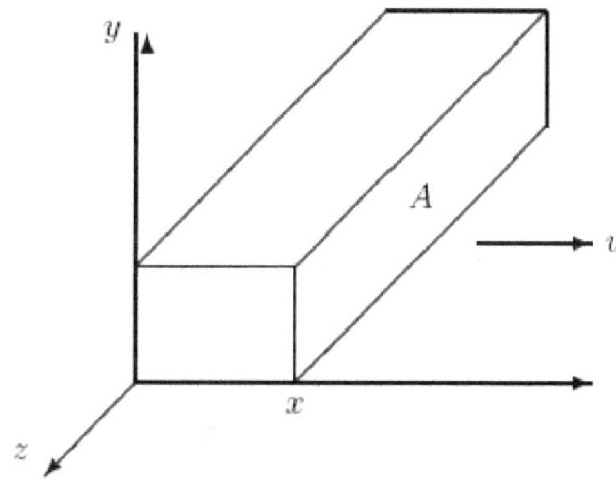

Figure 1: Packet of air moving with speed u

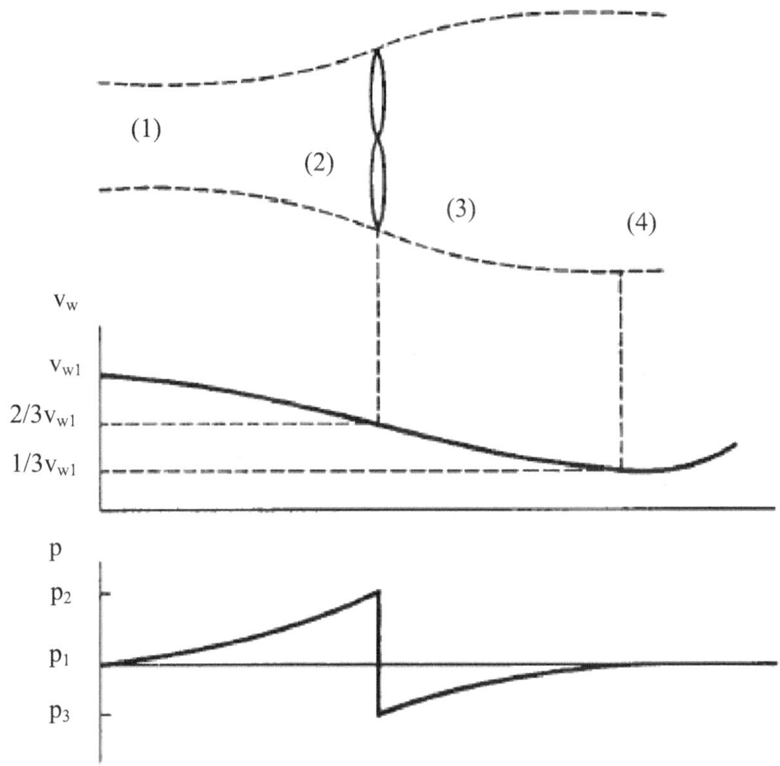

Fig. 2.2 Circular tube of air flowing through an ideal wind turbine [116]

This can be viewed as the power being supplied at the origin to cause the energy of the parcel to increase according to Eq. 2.1. A wind turbine will extract power from side, x, with Eq. 2.2 representing the total power available at this surface for possible extraction.

The physical presence of a wind turbine in a large moving air mass modifies the local air speed and pressure, as shown in Fig. 2.2. The picture is drawn for a conventional horizontal axis propeller type turbine.

Consider a tube of moving air with initial or undisturbed diameter, d_1, speed, v_{w1}, and pressure, p_1, as it approaches the turbine. The speed of the air decreases as the turbine is approached, causing the tube of air to enlarge to the turbine diameter, d_2. The air pressure will rise to the maximum just in front of the turbine and will drop below atmospheric pressure behind the turbine. Part of the kinetic energy in the air is converted to potential energy to produce this increase in pressure. Still more kinetic energy will be converted to potential energy after the turbine, to raise the air pressure back to atmospheric. This causes the wind speed to continue to decrease until the pressure is in equilibrium. Once the low point of wind speed is reached, the speed of the tube of air will increase back to $v_{w4} = v_{w1}$ as it receives kinetic energy from the surrounding air [117].

It can be shown [118] that under optimum conditions, when the maximum power is being transferred from the tube of air to the turbine, the following relationships hold:

$$\left.\begin{aligned} v_{w2} &= v_{w3} = \frac{2}{3} v_{w1} \\ v_{w4} &= \frac{1}{3} v_{w1} \\ A_2 &= A_3 = \frac{3}{2} A_1 \\ A_4 &= 3A_1 \end{aligned}\right\} \tag{2.3}$$

The mechanical power extracted is then the difference between the input and output power in the wind:

$$P_{m,ideal} = P_1 - P_4 = \frac{1}{2}\rho(A_1 v_{w1}^3 - A_4 v_{w4}^3) = \frac{1}{2}\rho(\frac{8}{9} A_1 v_{w1}^3) \tag{2.4}$$

This states that 8/9 of the power in the original tube of air is extracted by an ideal turbine. This tube is smaller than the turbine, however, and this can lead to confusing results. The normal method of expressing this extracted power is in terms of the undisturbed wind speed, v_{w1}, and the turbine area, A_2. This method yields

$$P_{m,ideal} = \frac{1}{2}\rho\left[\frac{8}{9}\left(\frac{2}{3}A_2\right)v_{w1}^3\right] = \frac{1}{2}\rho\left(\frac{16}{27}A_2 v_{w1}^3\right) \tag{2.5}$$

The factor 16/27 = 0.593 is called the Betz coefficient. It shows that an actual turbine cannot extract more than 59.3 percent of the power in an undisturbed tube of air of the same area. In practice, the fraction of power extracted will always be less because of mechanical imperfections. A good fraction is 35 – 40 % of the power in the wind under optimum conditions, although fractions as high as 50 %

have been claimed. A turbine extracts 40 % of the power in the wind, is extracting about two-thirds of the amount that would be extracted by an ideal turbine. This is rather good, considering the aerodynamic problems of constantly changing wind speed and direction as well as the frictional loss due to blade surface roughness.

2.1.2 Power Output from Practical Turbines

The fraction of power extracted from the power in the wind by a practical wind turbine is usually given by the symbol Cp, standing for the coefficient of performance or power coefficient. Using this notation and dropping the subscripts of Eq. 2.5, the actual mechanical power output can be written as

$$P_m = C_p\left(\frac{1}{2}\rho A v_w^{\ 3}\right) = \frac{1}{2}\rho\pi R^2 v_w^{\ 3} C_p(\lambda, \beta) \tag{2.6}$$

where, R is the blade radius of the wind turbine (m), V_w is the wind speed (m/sec), and ρ is the air density (kg/m^3). The coefficient of performance is not constant, but varies with the wind speed, the rotational speed of the turbine, and turbine blade parameters such as angle of attack and pitch angle. Generally, it is said that power coefficient, Cp, is a function of tip speed ratio, λ, and blade pitch angle, β (deg). The tip speed ratio is defined as

$$\lambda = \frac{\omega_R R}{v_w} \tag{2.7}$$

where, ω_R is the mechanical angular velocity of the turbine rotor in rad/s, and V_w is the wind speed in m/s. The angular velocity ω_R is determined from the rotational speed, n (r/min) by the equation

$$\omega_R = \frac{2\pi n}{60} \tag{2.8}$$

2.2 Wind Turbine Generator System (WTGS)

A wind turbine generator system (WTGS) transforms the energy present in the blowing wind into electrical energy. As wind is highly variable resource that cannot be stored, operation of a WTGS must be done according to this feature. The general scheme of a WTGS is shown in Fig. 2.3.

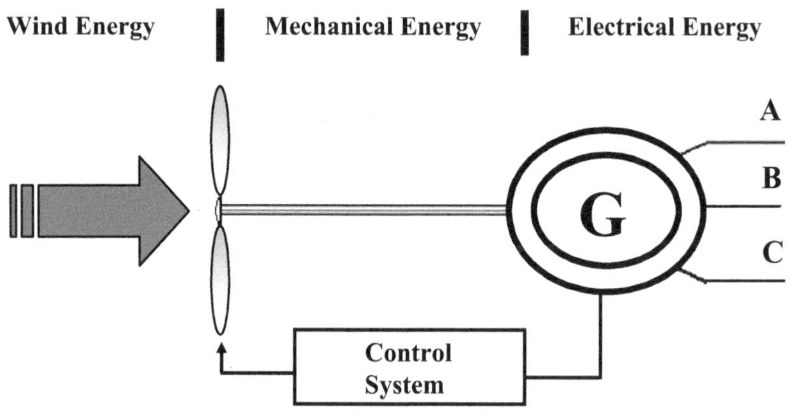

Fig. 2.3 General scheme of a WTGS where three types of energy states are presented: wind, mechanical, and electrical

A short overview of the system is given next. Wind energy is transformed into mechanical energy by a wind turbine that has several blades. It usually includes a gearbox that matches the turbine low speed to the higher speed of the generator. Some turbines include a blade pitch angle control (explained in Chap. 3) for controlling the amount of power to be transformed. Wind speed is measured with an anemometer.

The electrical generator transforms mechanical energy into electrical energy. Commercially available wind generators installed at present are squirrel cage induction generator, doubly fed induction generator, wound field synchronous generator (WFSG), and permanent magnet synchronous generator (PMSG). Based on rotational speed, in general, the wind turbine generator systems can be split into two types:

- Fixed speed WTGS
- Variable speed WTGS

2.3 Fixed Speed WTGS

A fixed speed WTGS consists of a conventional, directly grid coupled squirrel cage induction generator, which has some superior characteristics such as brushless and rugged construction, low cost, maintenance free, and operational simplicity. The slip and hence the rotor speed of a squirrel cage induction generator varies with the amount of power generated. These rotor speed variations are, however, very small, approximately 1 to 2 % of the rated speed. Therefore, this type of wind energy conversion system is normally referred to as a constant or fixed speed WTGS. The advantage of a constant speed system is that it is relatively simple. Therefore, the list price of constant speed turbines tends to be lower than that of variable speed turbines. However, constant speed turbines must be more mechanically robust than variable speed turbines [119]. Because the rotor speed cannot be varied, fluctuations in wind speed translate directly into drive train torque fluctuations, causing higher structural loads than with variable speed operation. This partly cancels the cost reduction achieved by using a relatively cheap generating system.

2.3.1 Fixed Speed WTGS Topology

The fixed speed WTGS topology is shown in Fig. 2.4. In the fixed speed WTGS topology, both single and double cage squirrel cage induction generators are commercially used. Note that squirrel cage induction generators used in wind turbines can often run at two different (but constant) speeds by changing the number of pole pairs of the stator winding. The relation between pole pairs and rotational speed is as follows:

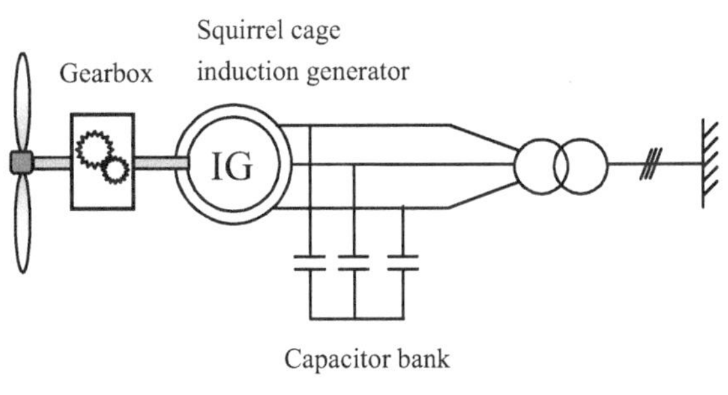

Fig. 2.4 Schematic diagram of a fixed speed WTGS

$$\omega_s = \frac{120f}{p} \tag{2.9}$$

where, f is the frequency of the stator voltage, p is the number of pole pairs, and ω_s is the generator rotor speed (rpm). There is a big speed difference between the turbine hub and the squirrel cage induction generator. Therefore, a gearbox is used in this topology that matches the turbine low speed to the higher speed of the generator.

A squirrel cage induction generator has a principal characteristic that it always consumes reactive power. To establish the rotating magnetic field of the stator, reactive power must be supplied from the network to the stator winding of the induction generator. In most cases, this is undesirable, particularly for large turbines and weak grids. Therefore, the reactive power consumption of the squirrel cage induction generator is nearly always partly or fully compensated by a capacitor bank to achieve a power factor close to one in the steady state. As a result, during load flow calculation this reactive power, Q_C, supplied by the capacitor bank should be considered. Figure 2.5 shows the schematic diagram of the entire power system including a wind turbine generator system (WTGS) that can be used for load flow study.

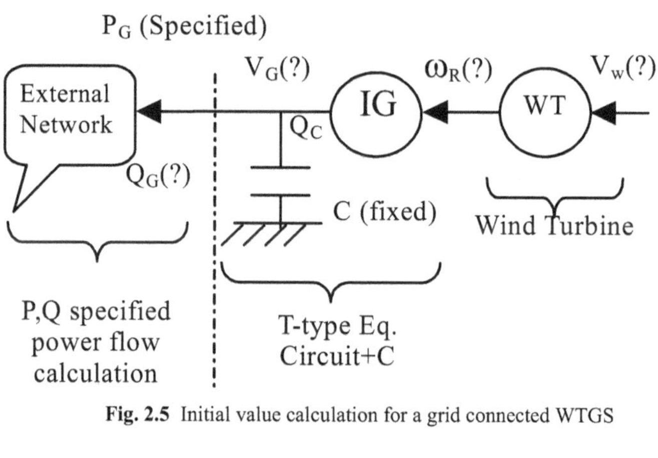

Fig. 2.5 Initial value calculation for a grid connected WTGS

The flowchart for determining the initial values of a WTGS connected to an external network is shown in Fig. 2.6. First, it is assumed that real power, P_G, and reactive power, Q_G, (initially zero) at the terminal of the IG are specified. By using these specified values, the entire power flow calculation should be done. Then a particular terminal voltage, V_G, is obtained. Next, the WTGS-SUBROUTINE, which also includes wind turbine characteristic, is called. Here, for specified V_G and P_G, wind speed, V_W, and slip, s, of the induction generator are determined from which the IG rotor speed, ω_R, can be calculated easily. Q_G is also determined within this subroutine. At this stage, the convergence of Q_G is checked. If Q_G does not converge, then the power flow calculation must be continued as shown in Fig. 2.6.

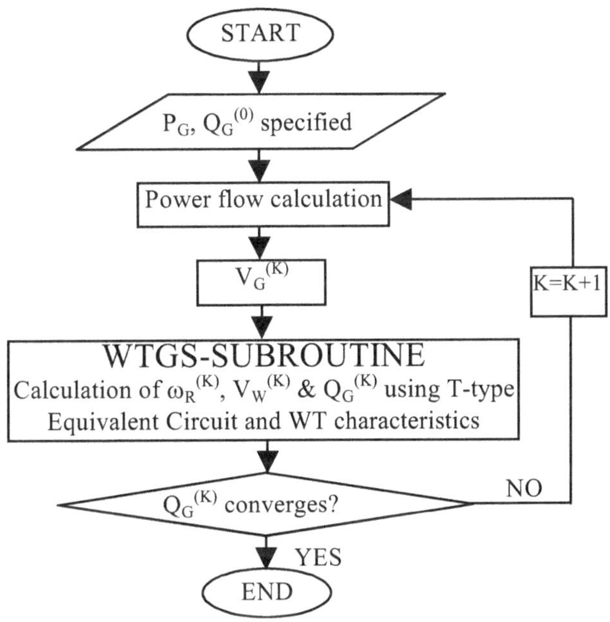

Fig. 2.6 Flowchart for power flow calculation including a WTGS

2.3.1.1 Equivalent Circuit Analysis

Both single and double cage induction generators are used as fixed-speed wind generators. The equivalent circuits of single and double cage induction generators are shown in Figs. 2.7 and 2.8, respectively, where s denotes rotational slip. From the single cage equivalent circuit of an IG shown in Fig. 2.7, the loop equations can be derived as Eqs. 2.10 and 2.11. From these two equations we can obtain the desired currents i_1 and i_2. Again, from the equivalent circuit of a single cage IG,

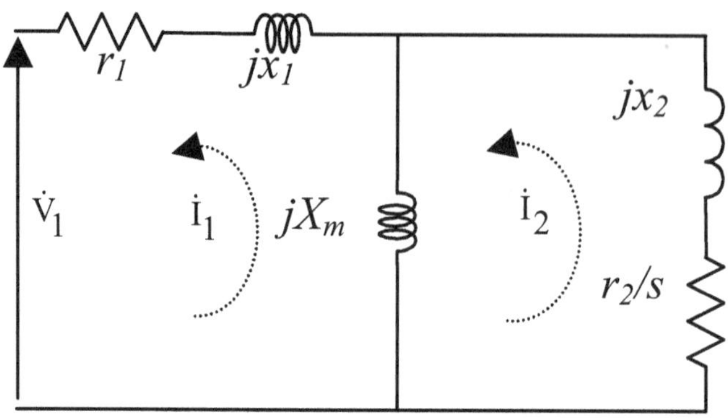

Fig. 2.7 Equivalent circuit of a single cage induction generator

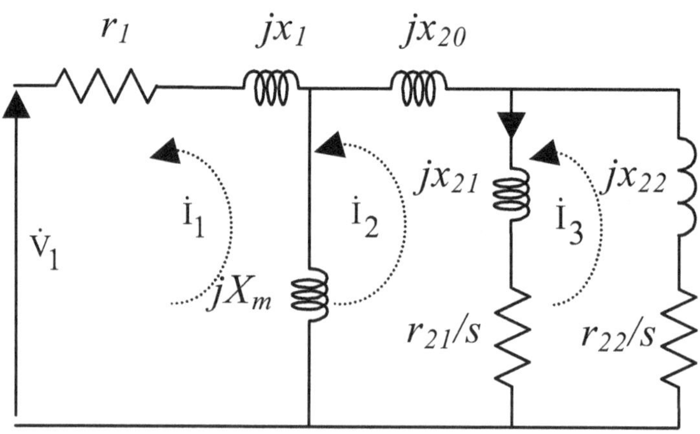

Fig. 2.8 Equivalent circuit of a double cage induction generator

we can calculate the input power of the induction generator, $P_{IG\_IN\_SINGLE}$, which is actually the output power of the wind turbine, shown in Eq. 2.12. From double cage equivalent circuit of the induction generator shown in Fig. 2.8, the loop equations can be derived as Eqs. 2.10, 2.13, and 2.14. From these three equations we can get the desired currents $\dot{I}_1$, $\dot{I}_2$, and $\dot{I}_3$. From the equivalent circuit of the double cage induction generator, we can calculate the input power of induction generator, $P_{IG\_IN\_DOUBLE}$, shown in Eq. 2.15. We can also calculate the desired output power of both single and double cage induction generators, P_{IG_OUT}, shown in Eq. 2.16, from the equivalent circuits shown in Figs. 2.7 and 2.8. The detailed description of the WTGS subroutine is presented below, in the light of the single cage equivalent circuit of the induction generator [43,108].

$$\dot{V}_1 = (r_1 + jx_1 + jx_m)\dot{I}_1 + jx_m\dot{I}_2 \tag{2.10}$$

$$0 = jx_m\dot{I}_1 - (\frac{r_2}{s} + jx_2 + jx_m)\dot{I}_2 \tag{2.11}$$

$$P_{IG\_IN\_SINGLE} = I_2^2 \frac{(1-s)}{s} r_2 \tag{2.12}$$

$$0 = jx_m\dot{I}_1 - (\frac{r_{21}}{s} + jx_{21} + jx_{20} + jx_m)\dot{I}_2 + (\frac{r_{21}}{s} + jx_{21})\dot{I}_3 \tag{2.13}$$

$$0 = (\frac{r_{21}}{s} + jx_{21})\dot{I}_2 - (\frac{r_{21}}{s} + \frac{r_{22}}{s} + jx_{21} + jx_{22})\dot{I}_3 \tag{2.14}$$

$$P_{IG\_IN\_DOUBLE} = I_3^2 \frac{(1-s)}{s} r_{22} + (I_3 - I_2)^2 \frac{(1-s)}{s} r_{21} \tag{2.15}$$

$$P_{IG\_OUT} = \mathrm{Re}\left[\dot{V}_1 \dot{I}_1^* \right] \tag{2.16}$$

2.3.1.2 WTGS-SUBROUTINE

Two probable cases have been considered for calculating the initial conditions of a wind turbine generator system.

Case I: In this case, it has been considered that the wind speed, V_W, is less than or equal to the rated wind speed. Therefore, the pitch angle, β, is set to zero. In order to get a desired output power from an induction generator, we need to know the exact wind speed. The whole procedure is briefly described in Fig. 2.9. First, we need to set two inputs. One is desired the IG output power, $P_{IG\text{-}OUT}$, and the other is the IG terminal voltage, V_T, that can be obtained from the power flow calculation. The initial values of the wind speed, V_W, and rotor speed, ω_R, are also needed. Now turbine torque, T_W, can be calculated by dividing the turbine mechanical power, P_m, shown in Eq. 2.6 by the tip speed ratio, λ, for $\beta = 0$.

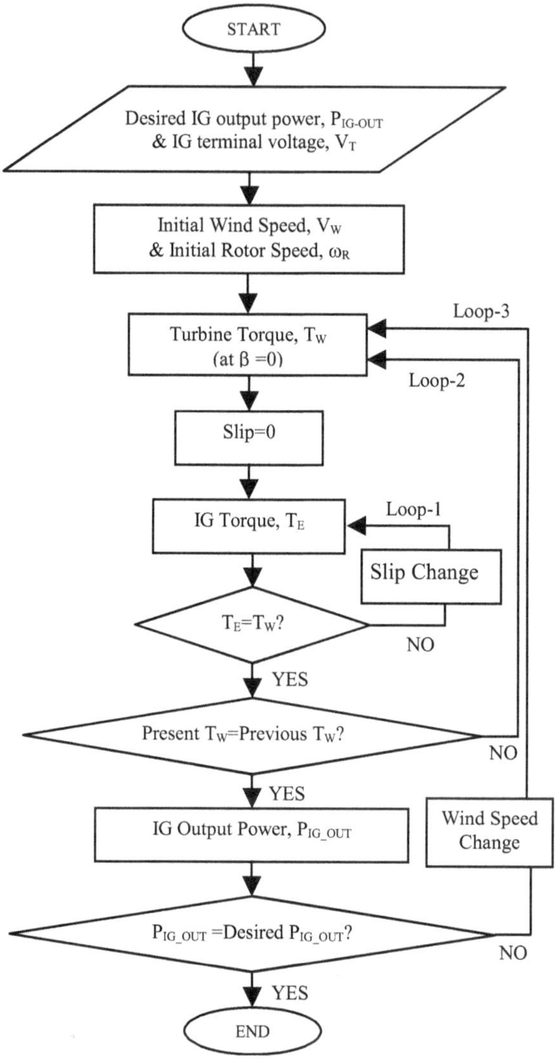

Fig. 2.9 Flowchart for initial value calculation (Case I)

The developed torque, T_E, of the IG can also be calculated from Eq. 2.12. The slip, s, is changed in loop-1 until T_E becomes equal to T_W. Using this slip, the new rotor speed, ω_R, can be calculated easily and this ω_R is used to calculate the wind turbine torque, T_W, instead of the initial ω_R until the present T_W becomes equal to the previous T_W in loop-2. When these two become equal, the present IG output power, $P_{IG\text{-}OUT}$, is calculated from Eq. 2.16. Loop-3 will be continued by changing the wind speed as shown in Fig. 2.9, until the present $P_{IG\text{-}OUT}$ becomes equal to the desired $P_{IG\text{-}OUT}$. Finally, we will get the desired the wind speed for any particular induction generator output power

Case II: In this case, it is considered that the wind speed, V_W, is above the rated speed. Therefore, we need to increase the pitch angle, β, to generate the rated induction generator output power. This process has been demonstrated in Fig. 2.10. Here, we assumed that wind speed, V_W, and terminal voltage, V_T, are known. The pitch angle, β, has been set to zero initially, and the rotor speed, ω_R, has also been approximated. Slip searching and the rotor speed determination process, which are already described in Case I, have been shown by dotted lines in Fig. 2.10. For a particular slip, when the present T_W and the previous T_W become equal, the IG output power, $P_{IG\text{-}OUT}$, is calculated. If it exceeds the rated output, then β is increased, and loop-3 will be continued until $P_{IG\text{-}OUT}$ becomes equal to the rated power. Therefore, finally, we will get a particular β for any wind speed over the rated speed, at which the induction generator output power becomes the rated power.

2.3.2 Fixed Speed Wind Turbine Characteristics

The modeling of a wind turbine rotor is somewhat complicated. According to the blade element theory [120], the modeling of blade and shaft needs complicated and lengthy computations. Detailed and accurate information about rotor geometry are also needed. For that reason, considering only the electrical behavior of the system, a simplified method of modeling the wind turbine blade and shaft is normally used.

For fixed speed wind turbine characteristics, the following Cp equations have been used from [33].

$$\lambda = \frac{V_W}{\omega_R} \tag{2.17}$$

$$C_p = \frac{1}{2}(\lambda - 0.022\beta^2 - 5.6)e^{-0.17\lambda} \tag{2.18}$$

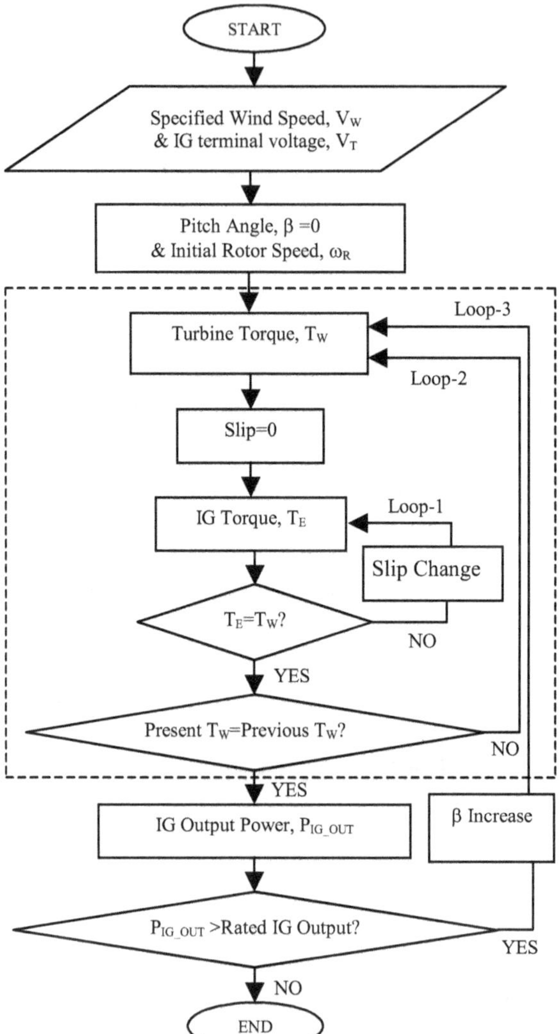

Fig. 2.10 Flowchart for initial value calculation (Case II)

In Eq. 2.17, the wind speed, V_w is in mile/hr. Keeping the original definition of tip speed ratio, λ, in Eq. 2.7, and changing the unit of wind speed from mile/hr to m/sec, Eqs. 2.17 and 2.18 are rearranged as shown below:

$$\lambda_i = \frac{3600R}{1609\lambda} \tag{2.19}$$

$$C_p = \frac{1}{2}(\lambda_i - 0.022\beta^2 - 5.6)e^{-0.17\lambda_i} \tag{2.20}$$

The Cp-λ curves for MOD2 wind turbine [33] are shown in Fig. 2.11 for different values of β. Power versus wind speed and pitch angle versus wind speed curves are shown together in Fig. 2.12.

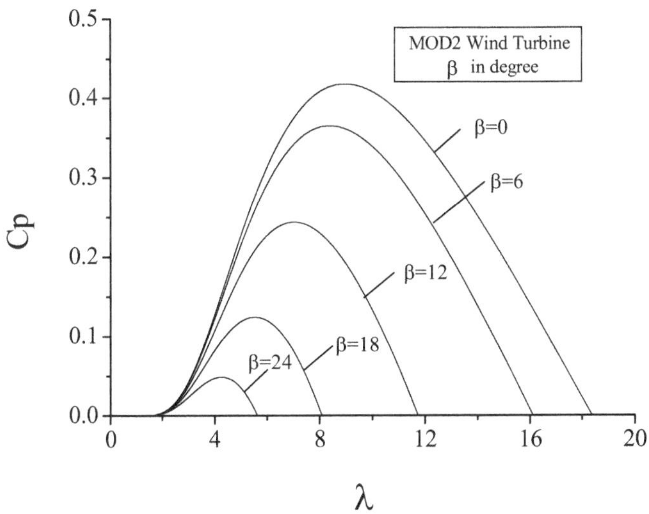

Fig. 2.11 Cp- λ curves for different pitch angles

2.3.3. Drive Train Modeling

The following four types of drive train models of the WTGS are usually available in the power system analysis:

- Six-mass drive train model
- Three-mass drive train model
- Two-mass shaft model
- One-mass or lumped model

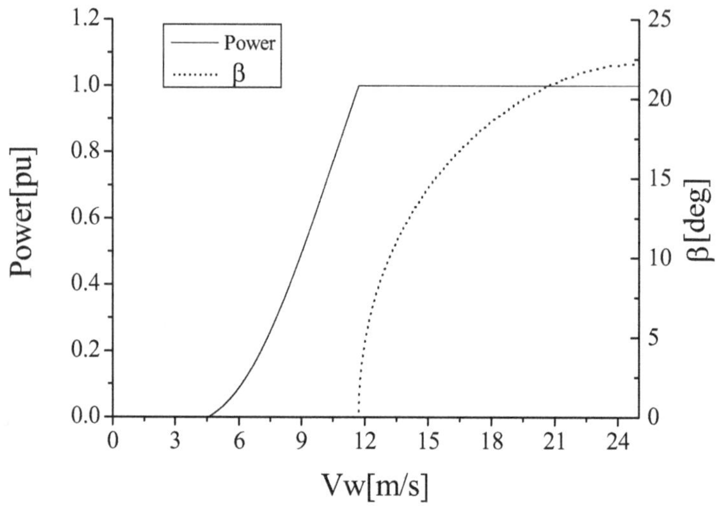

Fig. 2.12 Power vs. wind speed and pitch angle vs. wind speed characteristics

Figures 2.13a, b, d, and e show the above-mentioned six-mass, three-mass, two-mass, and one-mass drive train models, respectively. In the following few sections, the transformed three-mass drive train model shown in Fig. 2.13c is explained.

2.3.3.1 Six-Mass Drive Train Model

The basic six-mass drive train model is presented in Fig. 2.13a. The six-mass model system has six inertias: three blade inertias (J_{B1}, J_{B2}, and J_{B3}), hub inertia, J_H, gearbox inertia, J_{GB}, and generator inertia, J_G. θ_{B1}, θ_{B2}, θ_{B3}, θ_H, θ_{GB}, and θ_G represent angular positions of the blades, hub, gearbox and generator. ω_{B1}, ω_{B2}, ω_{B3}, ω_H, ω_{GB}, and ω_G correspond to the angular velocities of the blades, hub, gearbox, and generator. The elasticity between adjacent masses is expressed by the spring constants K_{HB1}, K_{HB2}, K_{HB3}, K_{HGB}, and K_{GBG}. The mutual damping between adjacent masses is expressed by d_{HB1}, d_{HB2}, d_{HB3}, d_{HGB}, and d_{GBG}. There exist some torque losses through external damping elements of individual masses, represented by D_{B1}, D_{B2}, D_{B3}, D_H, D_{GB}, and D_G. The model system needs generator torque, T_E and three individual aerodynamic torques acting on each blade (T_{B1},T_{B2}, and T_{B3}). The sum of the blade torques is the turbine torque, T_{WT}. It is assumed that the aerodynamic torques acting on the hub and gearbox are zero.

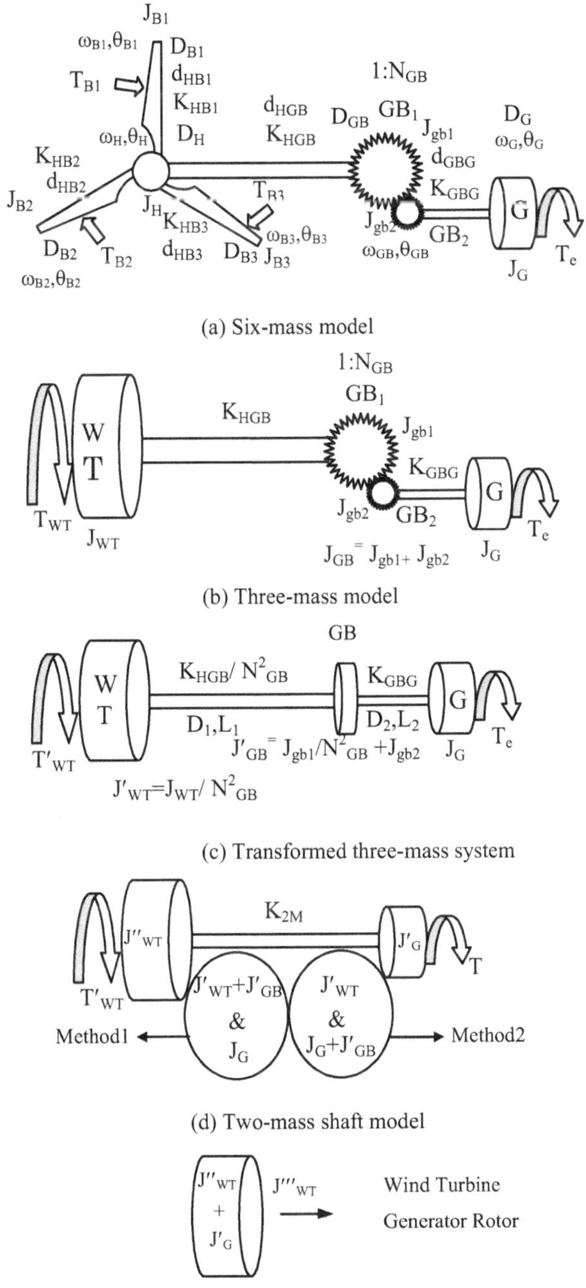

Fig. 2.13 Drive train models of wind turbine generator systems

2.3.3.2 Three-Mass Drive Train Model

The basic three-mass model is shown in Fig. 2.13b. The turbine inertia can be calculated from the combined weight of the three blades and hub. Therefore, the mutual damping between the hub and the blades is ignored in the three-mass model. Individual blade torque sharing cannot be considered in this model. Instead, it is assumed that the three-blade turbine has uniform weight distribution for simplicity, i.e., the turbine torque, T_{WT}, is assumed to be equal to the sum of the torque acting on the three blades. Therefore, the turbine can be looked upon as a large disk with small thickness. If proper data are not available, the simple equation below can be used for estimating the mass moment of inertia of a disk with small thickness [121].

$$J(Kg.m^2) = \frac{MD_d^2}{8} \tag{2.21}$$

where, D_d is the diameter of the disk and M is the weight of the disk. Similarly, generator and gearbox inertia can be calculated approximately from their diameter and weight. If we need precise estimation, precise data of geometry and very complicated formulas are needed for calculating the moment of inertia of the turbine, gearbox, and generator.

The shaft stiffness can be calculated from the equation below [122]:

$$K(Nm/rad) = \frac{G\pi D_{sh}^4}{32L} \tag{2.22}$$

where, D_{sh} is the shaft diameter, L is the shaft length, and G is the shear modulus. Stainless steel or ductile cast iron is normally used as the shaft material.

2.3.3.3 Geared System Transformation

When a torsional system is interconnected by a set of gears, the inertia disks are not being operated at the same angular speed throughout the system. In that case, the actual system needs to be corrected for the differences in the speeds of the component parts, i.e., the inertias and spring constants are referred to one speed of rotation, as shown in Fig. 2.13c. The basis for these transformations is that the potential and kinetic energies of the equivalent system should be the same as those of the actual one, and it is assumed that the gear teeth do not break contact while transmitting vibration. The above-mentioned transformation can be summarized as follows [122, 123]:

$$\frac{J_{eq}}{J_a} = \frac{K_{eq}}{K_a} = (\text{speed ratio})^2 \tag{2.23}$$

where, the suffixes 'eq' and 'a' means equivalent and actual, respectively.

2.3.3.4 Two-Mass Drive Train Model

The three-mass system can be converted into a two-mass system, which is shown in Fig. 2.13d by adding the masses of two disks together and by connecting the two disks with equivalent shaft stiffness. The equivalent shaft stiffness of the two-mass system, K_{2M}, can be determined from the parallel shaft stiffness as in Eq. 2.24 [122, 123]. In Fig. 2.13d, J''_{WT} and J'_G represent the equivalent mass moments of inertia of the wind turbine and generator, respectively.

Note here that the two disks should be added together by considering the lower shaft stiffness. For example, if the spring constant of the low-speed side is lower than that of the high-speed side, then the gearbox and generator inertias should be added, as shown in method-2 of Fig. 2.13d and vice versa, which can be ensured from the simulation results presented in Sect. 2.3.4.3.1. This might be a good practice for wind turbine drive train conversion methodology instead of the conventional one of connecting the turbine and gearbox together, as presented in [26]. Accordingly, the self damping of the generator and gearbox should be added together and the mutual damping of the gearbox and generator is neglected in the two-mass shaft model:

$$\frac{1}{K_{2M}} = \frac{1}{K_{HGB} / N_{GB}^2} + \frac{1}{K_{GBG}} \tag{2.24}$$

2.3.3.5 One-Mass Lumped Model

In the one-mass or lumped model, all types of windmill drive train components are lumped together and work as a single rotating mass, as shown in Fig. 2.13e. The dynamic behavior can be expressed by the following differential equation:

$$\frac{d\omega_R}{dt} = \frac{T_{WT} - T_E}{J'''_{WT}} \tag{2.25}$$

where, J'''_{WT} is the inertia constant of the rotating mass, ω_R is the rotor speed, T_{WT} is the input mechanical torque applied to the wind turbine rotor and T_E is the electromagnetic torque of the induction generator.

2.3.3.6 Per Unitization

In the drive train model system, all data used in the state equations are converted to a per unit system. If P_B is the base power (VA), ω_0 the base electrical angular velocity (rad/sec) and P the number of pole pairs of the generator, the base values of the per unit system at the high-speed side of the drive train are defined as follows:

The base mechanical speed (mech. rad/sec), $\omega'_B = \omega_0/P$

The base torque (Nm), $T'_B = P_B/\omega'_B$

The base inertia [Nm/(rad/sec)], $J'_B = \dfrac{T'_B}{0.5\omega'_B} = \dfrac{P_B}{0.5\omega'^2_B}$

The base spring constant, [Nm/(rad/sec)], $K'_B = \dfrac{T'_B}{\omega'_B} = \dfrac{P_B}{\omega'^2_B}$

The base damping constant, [Nm/(rad/sec)], $D'_B = d'_B = \dfrac{T_B}{\omega'_B} = \dfrac{P_B}{\omega'^2_B}$

Now, the low-speed side (turbine-side) base quantities can be calculated from the high-speed side (generator-side) base quantities using the gearbox speed ratio, N_{GB}, as follows:

$$\begin{aligned} \omega''_B &= \omega'_B / N_{GB} & J''_B &= N^2_{GB} J'_B \\ \theta''_B &= \theta'_B / N_{GB} & D''_B &= N^2_{GB} D'_B \\ T''_B &= N_{GB} T'_B & K''_B &= N^2_{GB} K'_B \end{aligned} \tag{2.26}$$

In the simulation study, $H_{B(1,2,3)}$, H_H, H_{GB}, and H_G represent the per unit inertia constants (sec) of three blades, hub, gearbox, and generator, respectively.

2.3.3.7 Wind Farm Equivalent n-Machine

For the transient stability analysis of a WTGS, it is cumbersome to simulate each individual wind turbine. Therefore, wind turbines with the same torsional natural frequency might be added as follows [12, 27]:

$$\left.\begin{aligned} J_{wt} &- \sum_{i=1}^{p} J_{wti} \\ J_{gb} &= \sum_{i=1}^{p} J_{gbi} \\ J_{g} &= \sum_{i=1}^{p} J_{gi} \\ K &= \sum_{i=1}^{p} K_{i} \end{aligned}\right\} \tag{2.27}$$

where, i = number of each individual wind turbine and p = total number of wind turbine.

2.3.4 Comparative Study Among Different Types of Drive Train Modeling

2.3.4.1 Simulation Model

Two types of model systems are used for the sake of exact comparison among different types of drive train models of a WTGS. Figure 2.14 shows model system I, where one synchronous generator (SG) is connected to an infinite bus through a transformer and a double circuit transmission line. In the figure, the double circuit transmission line parameters are numerically shown in the form of R+jX, where R and X represent the resistance and reactance, respectively. One wind farm (Induction generator, IG) is connected with the network via a transformer and a short transmission line. A capacitor bank has been used for reactive power compensation at steady state. The value of capacitor C is chosen so that the power factor of the wind power station becomes unity [43] during the rated operation. Automatic voltage regulator (AVR) and governor (GOV) control system models shown in Figs. 2.15 and 2.16, respectively have been included in the synchronous generator model in the simulations. Figure 2.17 shows model system II, where the aggregated model of the wind farm (induction generator) is directly connected to the synchronous generator through a double circuit transmission line.

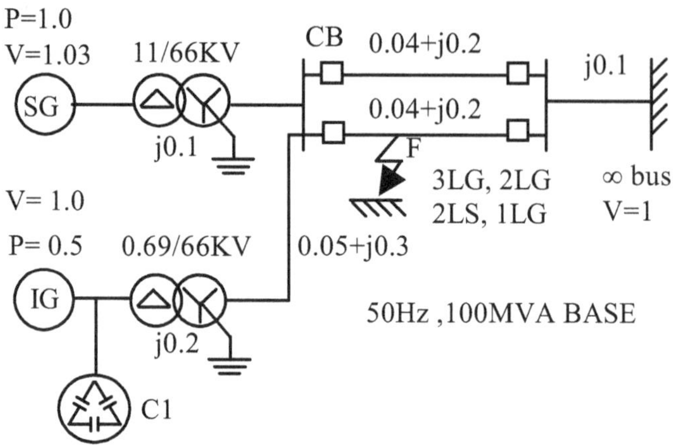

Fig. 2.14 Model system I

The IEEE generic turbine model and approximate mechanical-hydraulic speed governing system are used with a synchronous generator [124]. The IEEE alternator supplied rectifier excitation system (AC1A) [125] is used for excitation control of the synchronous generator. The generator parameters for both model systems are shown in Table 2.1. The initial values used for model systems I and II are shown in Tables 2.2 and 2.3, respectively. The drive train parameters for the six-mass, three-mass, and two-mass models are shown in Tables 2.4 and 2.5. For transient stability analysis, the symmetrical three-line-to-ground fault, 3LG, is considered. Some unsymmetrical faults such as double-line-to-ground fault, 2LG (phases a and b), line-to-line fault, 2LS (between phases a and b), and single-line-to-ground fault, 1LG (phase a) are also considered. Time step and simulation time were chosen 0.00005 sec and 10 sec, respectively. The simulations were done by using PSCAD/EMTDC[1] [126].

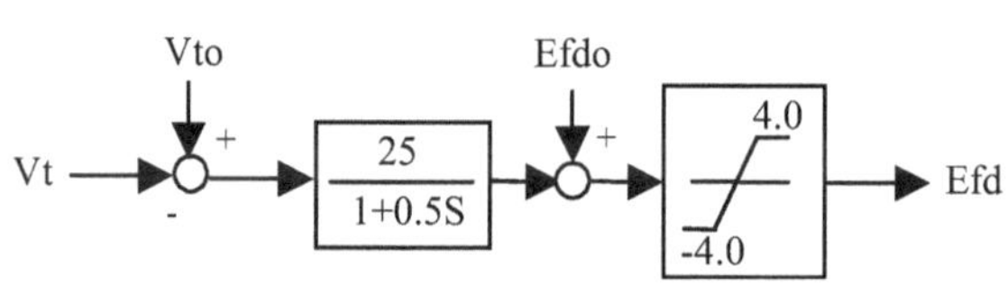

Fig. 2.15 AVR model

[1] For the latest information on PSCAD/EMTDC, visit at http://pscad.com

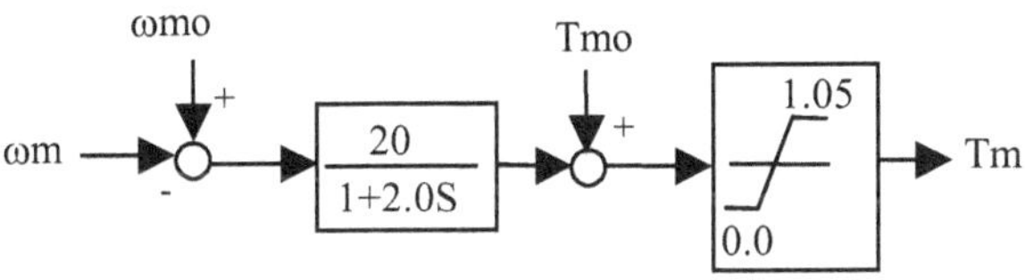

Fig. 2.16 GOV model

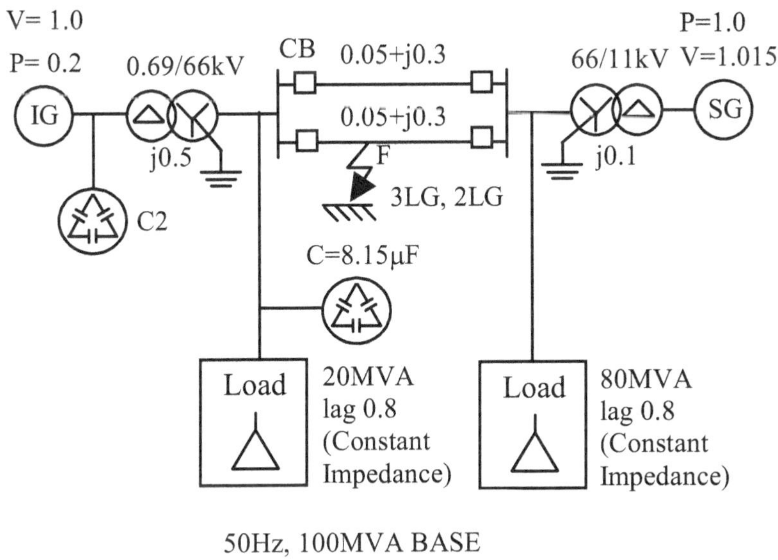

Fig. 2.17 Model system II

2.3.4.2 Effects of Equal and Unequal Blade Torque Sharing on Stability

Wind speed is intermittent and stochastic by nature. Therefore, the torques acting on the three blades of a wind turbine are not always equal. In this section, the effect of equal and unequal torque sharing on transient stability are analyzed using the six-mass drive train model. A 3LG fault is considered to occur at fault point F of model system I. The initial values for the stable and unstable cases are shown as Condition 1 and Condition 2 in Table 2.2. The drive train parameters of the six-mass model are shown in Table 2.4, but all types of damping are disregarded in this case to consider the worst-case scenario.

Table 2.1 Generator parameters

SG		IG	
MVA	100	MVA	50/20
ra (pu)	0.003	r1 (pu)	0.01
xa (pu)	0.13	x1 (pu)	0.1
Xd (pu)	1.2	Xmu (pu)	3.5
Xq (pu)	0.7	r21 (pu)	0.035
Xd′ (pu)	0.3	x21 (pu)	0.030
Xq′ (pu)	0.22	r22 (pu)	0.014
Xd″ (pu)	0.22	x22 (pu)	0.098
Xq″ (pu)	0.25		
Tdo′ (sec)	5.0		
Tdo″ (sec)	0.04		
Tqo″ (sec)	0.05		
H (sec)	2.5		

Table 2.2 Initial conditions of generators and turbines (model system I)

	Condition 1		**Condition 2**		**Condition 3**	
	SG	**IG**	**SG**	**IG**	**SG**	**IG**
P(pu)	1.0	0.39	1.0	0.40	1.0	0.50
V(pu)	1.03	1.042	1.03	1.039	1.03	0.999
Q(pu)	0.244	0.053 (0.206)*	0.251	0.049 (0.209)*	0.334	0.000 (0.239)*
Efd(pu)	1.719	-	1.725	-	1.803	-
Tm(pu)	1.003	-	1.003	-	1.003	-
SG_δ(deg)	50.47	-	50.50	-	50.72	-
slip	0.0	−0.769%	0.0	−0.794%	0.0	−1.09%
Vw (m/s)	-	10.615	-	10.722	-	11.797
β (deg)	-	0	-	0	-	0

* Reactive power drawn by an induction generator

Table 2.3 Initial conditions of generators and turbines (model system-II)

	Condition 4		**Condition 5**	
	SG	**IG**	**SG**	**IG**
P(pu)	0.624	0.16	0.58	0.20
V(pu)	1.015	1.014	1.015	1.00
Q(pu)	0.279	0.018 (0.081)*	0.297	0.000 (0.095)*
Efd(pu)	1.50	-	1.50	-
Tm(pu)	0.626	-	0.582	-
slip	0.0	−0.835%	0.0	−1.09%
Vw (m/s)	-	10.74	-	11.79
β (deg)	-	0	-	0

* Reactive power drawn by an induction generator

Table 2.4 Six-mass and three-mass drive train model parameters of a WTGS in per unit (from [32]) [based on high-speed rotation]

	6 M	3 M		6 M	3 M		6 M	3 M
$H_{B(1,2,3)}$	0.6388	-	K_{HGB}	54.75	54.75	D_G	0.01	0.01
H_H	0.0114	-	K_{GBG}	1834.1	1834.1	$d_{HB(1,2,3)}$	12.0	-
H_{WT}	-	1.9277	$D_{B(1,2,3)}$	0.004	-	d_{HGB}	3.5	3.5
H_{GB}	0.0806	0.0806	D_H	0.01	-	d_{GBG}	10.0	10.0
H_G	0.1419	0.1419	D_{WT}	-	0.022			
$K_{HB(1,2,3)}$	1259.8	-	D_{GB}	0.022	0.022			

Table 2.5 Two-mass drive train model parameters of a WTGS in per unit (transformation is based on six-mass drive train parameters)

2M					
H''_{WT}	1.9277	K_{2M}	53.16	D'_G	0.032
H'_G	0.2225	D''_{WT}	0.022	d''_{2M}	3.5

The responses of the IG rotor and turbine hub speeds are shown in Figs. 2.18 and 2.19, respectively, for both stable and unstable situations. It is seen from Figs. 2.18 and 2.19 that unequal blade torque sharing has no effect on the transient stability of a WTGS. Therefore, the three-mass and two-mass reduced order drive train models can be used, where it is assumed that the turbine torque is equal to the sum of the torques acting on the three blades. The comparison among the three types of drive train models of a WTGS during network disturbances is presented in the next section by using model systems I and II.

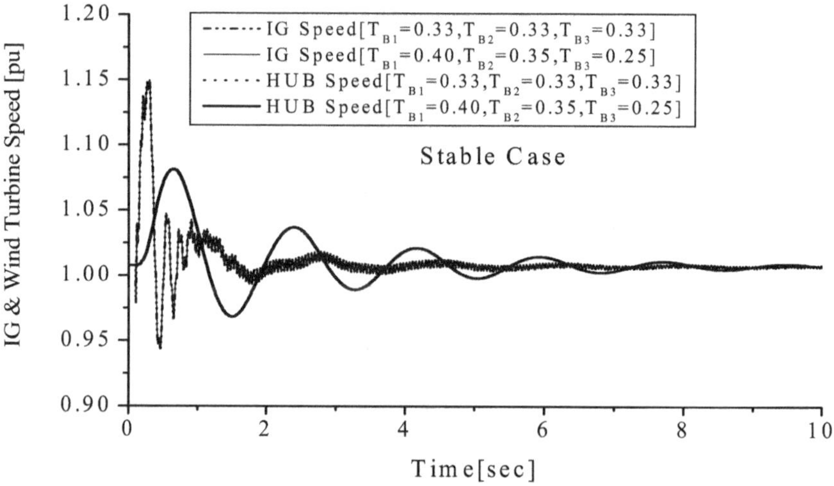

Fig. 2.18 Transient effect of equal and unequal blade torque sharing (Condition 1, 3LG, stable case, model system I)

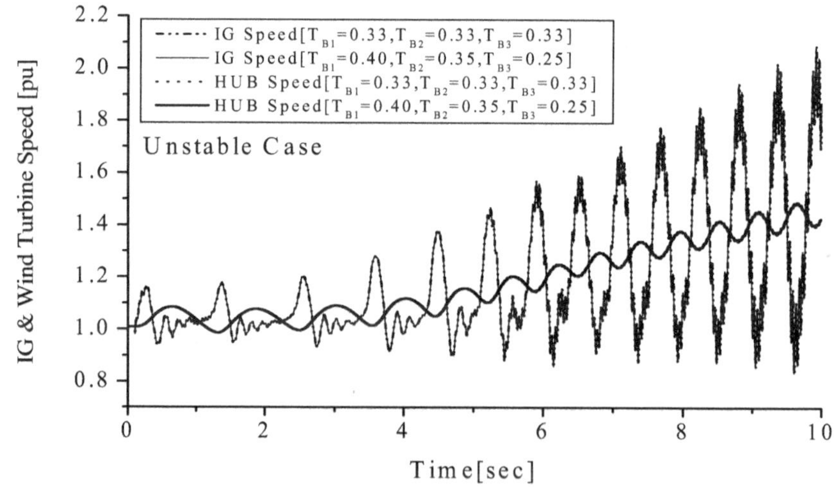

Fig. 2.19 Transient effect of equal and unequal blade torque sharing (Condition 2, 3LG, unstable case, model system I)

2.3.4.3 Comparison Using Model System I

First, the effect of the drive train parameters of the six, three, and two-mass drive train models on the transient stability of a WTGS are analyzed by using model system I. The comparison is carried out at different IG output power levels for different types of symmetrical and unsymmetrical faults. The fault is considered to occur at 0.1 sec, the circuit breakers (CB) on the faulted line are opened at 0.2 sec and at 1.0 sec, are reclosed.

2.3.4.3.1 Effects of Drive Train Parameters on the Transient Stability of WTGS

The effects of the drive train parameters on the transient stability of a grid connected WTGS are analyzed by considering the severe 3LG fault that occurred at point F of Fig. 2.14. The initial values are presented as Condition 2 in Table 2.2.

a. Effect of Inertia Constants

In this sub-section, the effects of inertia constant on the transient stability of a WTGS are analyzed using the six-mass, three-mass, and two-mass drive train models. All types of damping are neglected in this case. For the two-mass shaft model, two types of inertia sets are used, as shown in Fig. 2.13d. Responses of the IG speed and turbine speed are shown in Figs. 2.20 and 2.21, respectively for the six-mass, three-mass, and two-mass drive train models. Some other simulation results of IG rotor and turbine speeds are presented in Figs. 2.22 and 2.23, respectively, where the turbine inertia of each drive train model is increased by 50 % from the original value shown in Tables 2.4 and 2.5. In Figs. 2.24 and 2.25, the IG

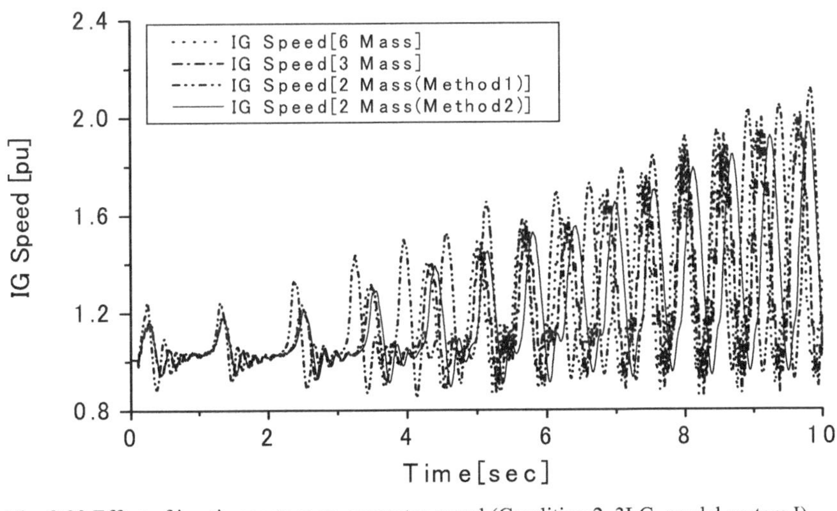

Fig. 2.20 Effect of inertia constant on generator speed (Condition 2, 3LG, model system I)

rotor and turbine speeds are shown, respectively, where the generator inertia constant of each drive train model is increased by about 50 % from the original value shown in Tables 2.4 and 2.5. It is clear from Figs. 2.24 and 2.25 that the transformation from three-mass to two-mass is needed to perform according to method 2 of Fig. 2.13d, i.e., the gearbox and generator masses should be added together as they are separated with comparatively lower shaft stiffness. Moreover, it is seen that the increase of turbine and generator inertia constants enhances the transient stability of the WTGS for all types of drive train models. From Figs. 2.20 – 2.25, it is clear that if the proper transformation process is applied, then the two-mass shaft model shows almost the same transient characteristics as those of the six-mass and three-mass drive train models.

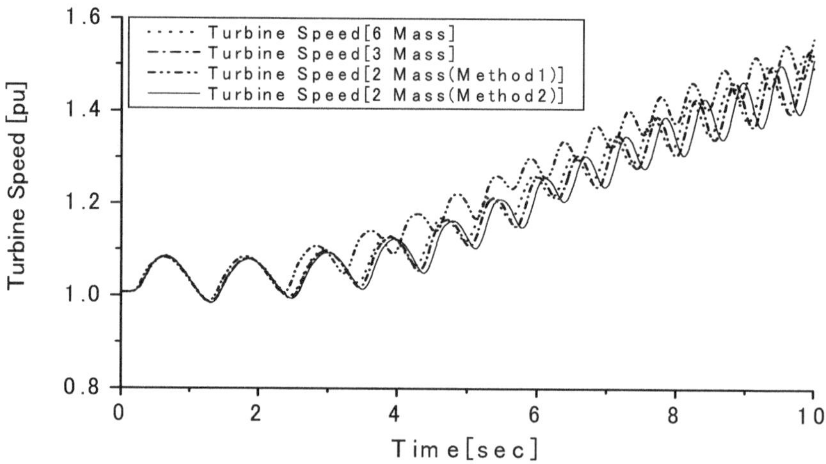

Fig. 2.21 Effect of the inertia constant on the turbine speed (Condition 2, 3LG, model system I)

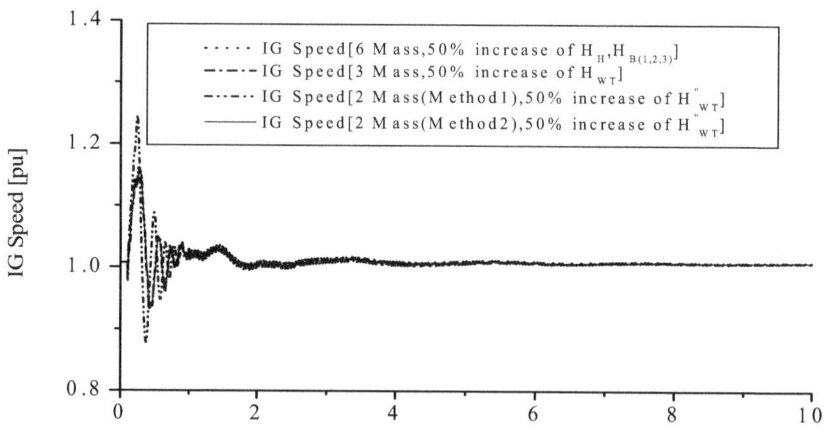

Fig. 2.22 Effect of the increased inertia constant of the turbine on the generator speed (Condition 2, 3LG, model system I)

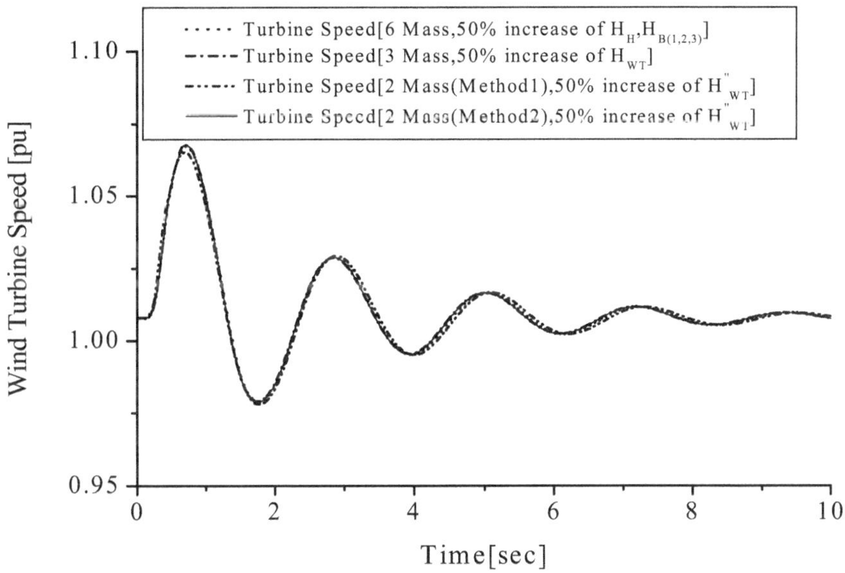

Fig. 2.23 Effect of the increased inertia constant of the turbine on the turbine speed (Condition 2, 3LG, model system I)

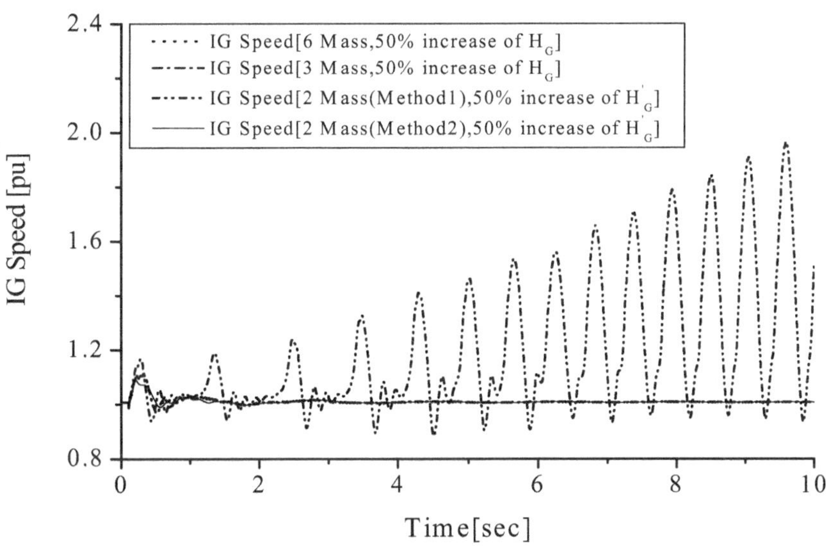

Fig. 2.24 Effect of the increased inertia constant of the generator on its rotor speed (Condition 2, 3LG, model system I)

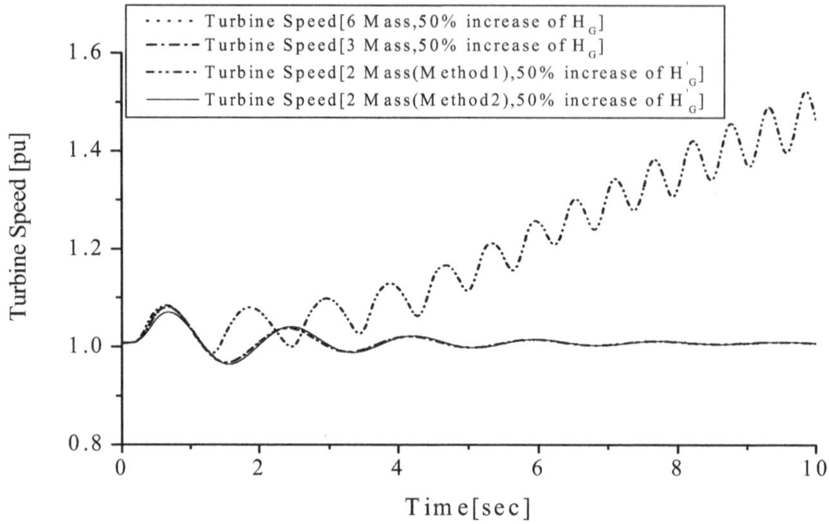

Fig. 2.25 Effect of the increased inertia constant of the generator on the turbine speed (Condition 2, 3LG, model system I)

b. Effect of Spring Constants

In this section, the effect of a spring constant on the transient stability of a WTGS is demonstrated using the six-mass, three-mass and two-mass drive train models. Here, the damping is neglected. The 3LG fault is considered to occur at point F of Fig. 2.14. First, the results of the transient characteristics of the six-mass drive

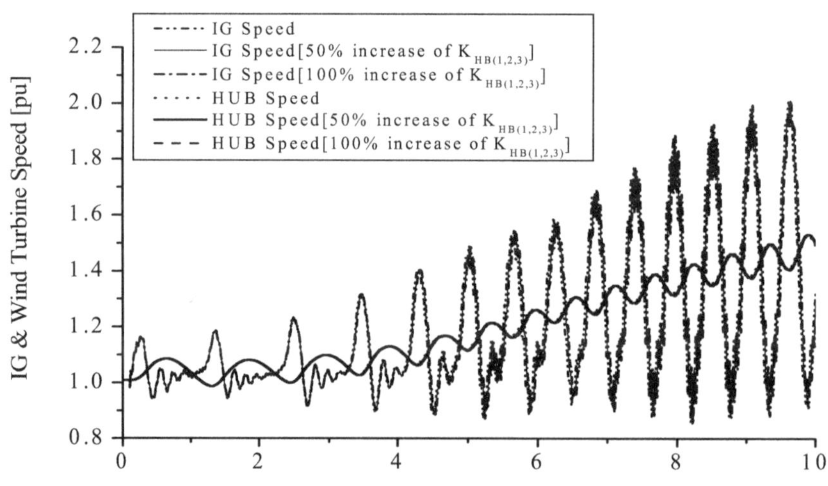

Fig. 2.26 Effect of the spring constant between the hub-blades of the six-mass model (Condition 2, 3LG, model system I)

train model are presented where the stiffness between the hub-blades, gearbox-generator, and hub-gearbox are increased by certain percentages from the original values as shown in Figs. 2.26, 2.27, and 2.28, respectively. It is seen from those figures that the stiffnesses between the hub-blades and gearbox-generator have a negligible effect on the transient stability of a WTGS. But the transient stability strongly depends on the spring constant between the hub-generator. In the three-mass drive train model, it is also seen that the spring constant between the

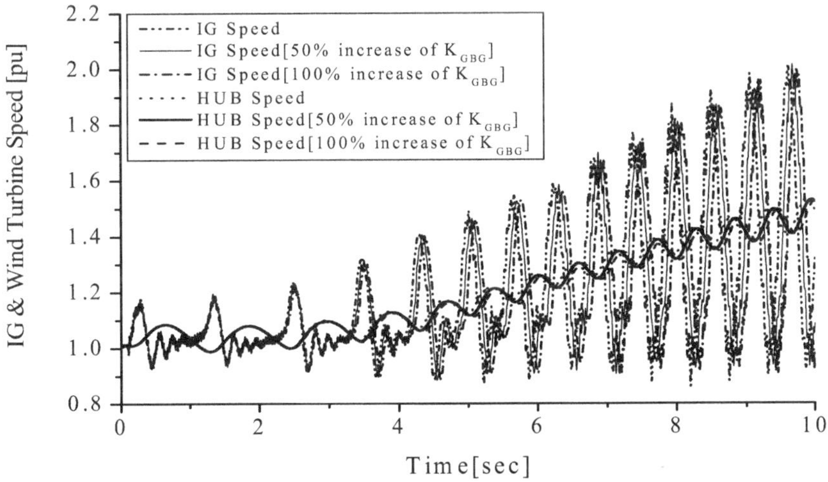

Fig. 2.27 Effect of the spring constant between the gearbox-generator of the six-mass model (Condition 2, 3LG, model system I)

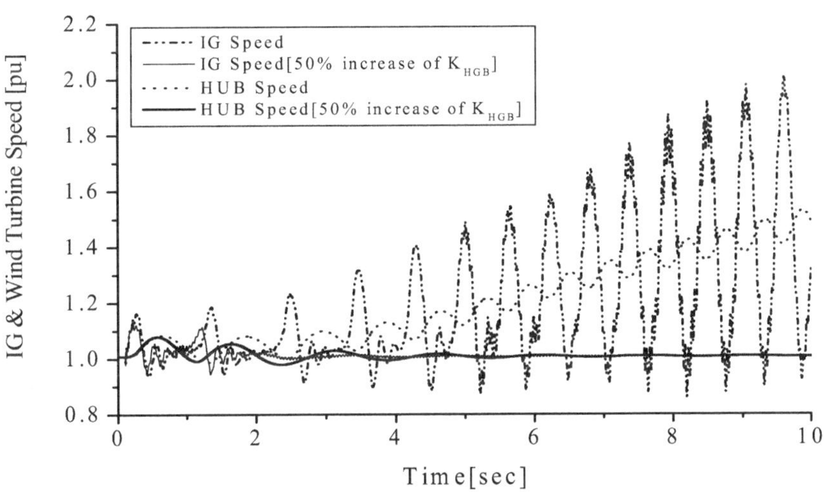

Fig. 2.28 Effect of the spring constant between the hub-gearbox of the six-mass model (Condition 2, 3LG, model system I)

gearbox-generator has almost no effect on stability, as shown in Fig. 2.29. Finally, the effect of the spring constant between the hub-gearbox for the six-mass and three-mass models and the effect of equivalent stiffness of the two-mass model are compared in Fig. 2.30 for a severe network disturbance in the model system. Moreover, these stiffnesses are increased by 50 % from the original values and their effects can be observed in Fig. 2.31. It is clear from Figs. 2.30 and 2.31 that the two-mass shaft model has almost the same transient characteristics as those of the six-mass and three-mass drive train models under a network disturbance.

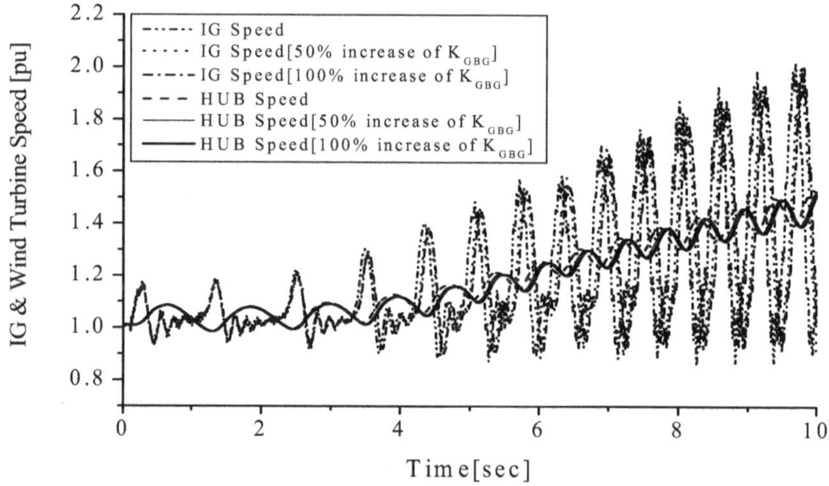

Fig. 2.29 Effect of the spring constant between the gearbox-generator of the three-mass model (Condition 2, 3LG, model system I)

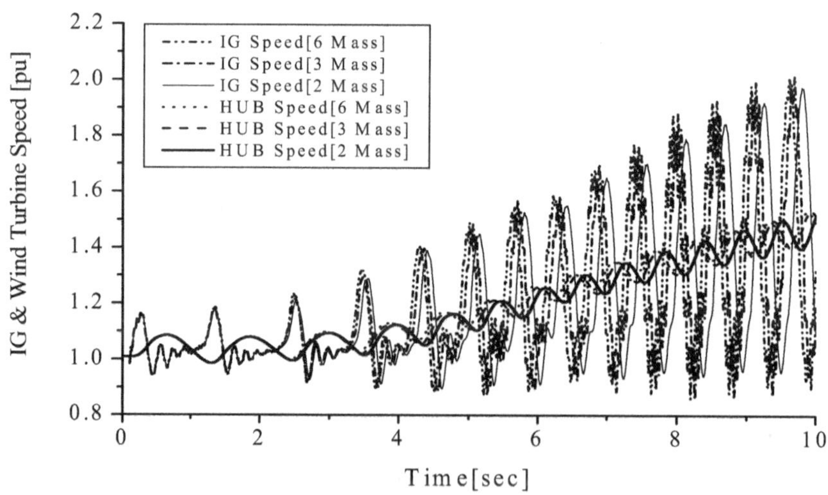

Fig. 2.30 Effect of the spring constant between the hub-gearbox of the six, three, and two mass models (Condition 2, 3LG, model system I)

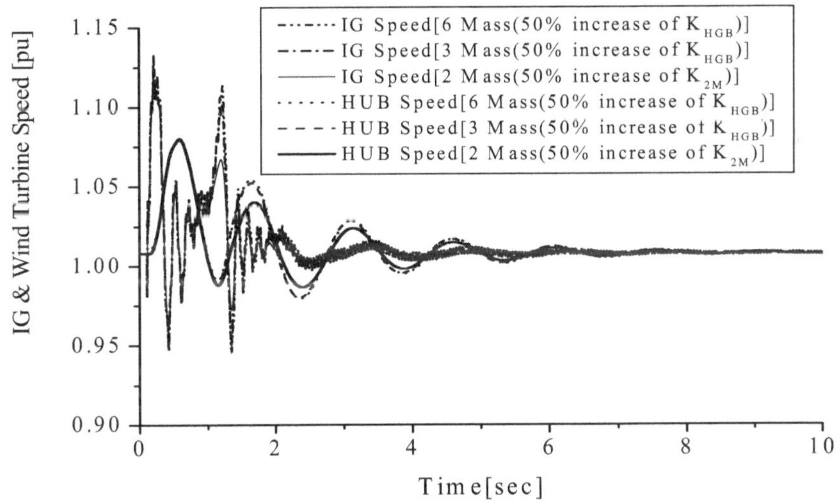

Fig. 2.31 Effect of the increased spring constant between the hub-gearbox of the six, three, and two mass models (Condition 2, 3LG, model system I)

c. Effect of Damping Constants

In this section, the effect of self and mutual damping of a drive train are analyzed using the six-mass, three-mass, and two-mass models under a severe network disturbance in the model system. The responses of the IG rotor and turbine speeds of the the six-mass model are shown in Figs. 2.32 and 2.33, respectively, where the damping constant is considered or disregarded. It is seen that both self and mutual

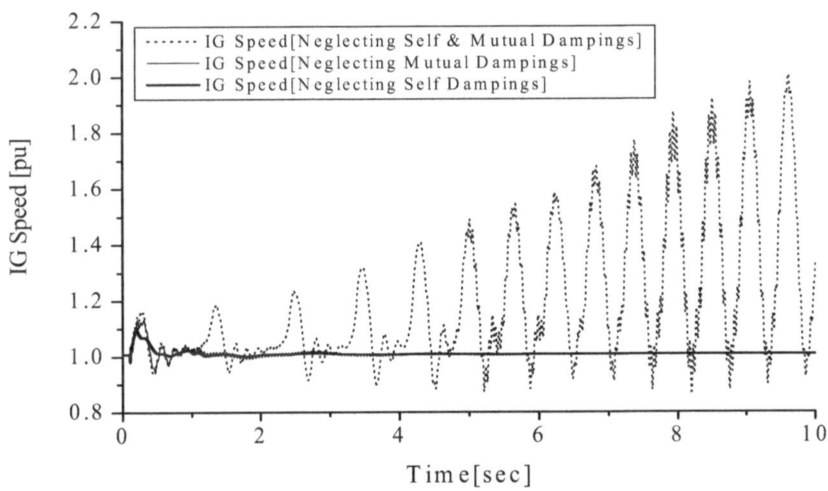

Fig. 2.32 Effect of the damping constants of the six-mass model on the generator speed (Condition 2, 3LG, model system I)

damping have significant effects on the transient stability of a WTGS, and among these two dampings, the mutual damping makes the WTGS transiently more stable. But in the three-mass model, it is not possible to consider the mutual damping between the hub and blades. In the two-mass model, only the mutual damping between the hub and gearbox is present. External damping elements represent torque

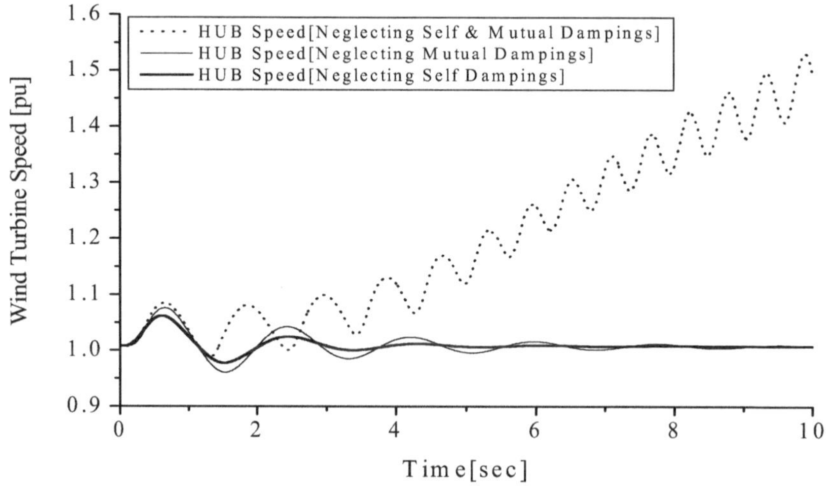

Fig. 2.33 Effect of the damping constants of the six-mass model on the turbine speed (Condition 2, 3LG, model system I)

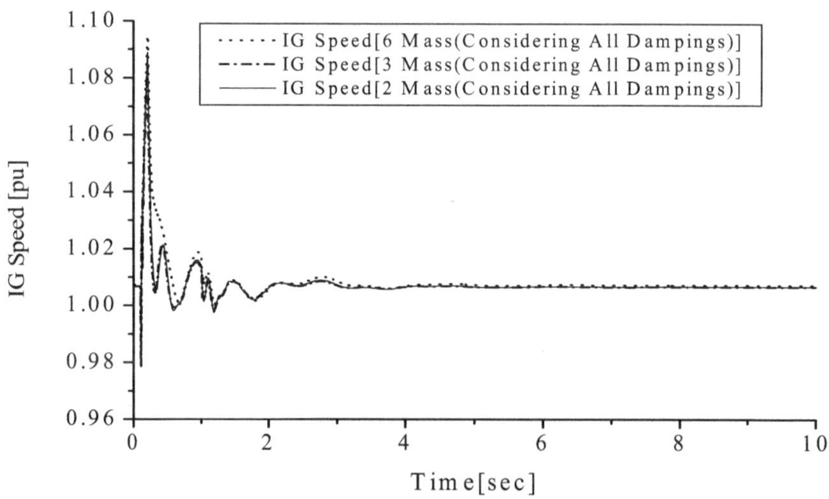

Fig. 2.34 Effect of the damping constants of the six, three, and two-mass models on the generator speed (Condition 2, 3LG, model system I)

losses. As the generator and gearbox masses are lumped together in the two-mass model, the self-damping of the individual elements is also lumped together. Finally, simulations have been carried out using the damping values shown in Table 2.4 and it is found that the two-mass, three-mass, and six-mass drive train models give almost the same results shown in Figs. 2.34 and 2.35.

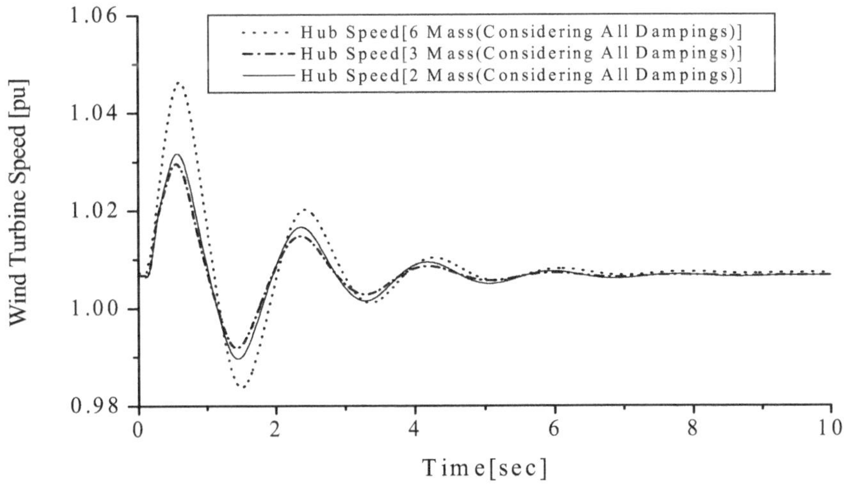

Fig. 2.35 Effect of the damping constants of the six, three, and two-mass models on the turbine speed (Condition 2, 3LG, model system I)

2.3.4.3.2 Fault Analysis

The transient stability of WTGS is analyzed again here using the six-mass, three-mass, and two-mass drive train models against different types of symmetrical and unsymmetrical faults in model system I. Drive train parameters are taken from Tables 2.4 and 2.5. The initial values at different IG output power levels can be obtained from the method described in [43]. The simulation results with and without considering the damping are shown briefly in Tables 2.6 and 2.7, where O and × represent stable and unstable situations of the WTGS, respectively. It is clear from the simulation results that in all cases the two-mass shaft model give the same results as those of the three-mass and six-mass drive train models.

Table 2.6 Transient stability results for two-mass, three-mass and six-mass models (neglecting all types of damping, Condition 3)

IG POWER (MW)	1LG fault			2LS fault			2LG fault			3LG fault		
	2M	3M	6M	2M	3M	6M	2M	3M	6M	2M	3M	6M
50	O	O	O	O	O	O	×	×	×	×	×	×
44	O	O	O	O	O	O	×	×	×	×	×	×
43	O	O	O	O	O	O	O	O	O	×	×	×
40	O	O	O	O	O	O	O	O	O	×	×	×
39	O	O	O	O	O	O	O	O	O	O	O	O

Table 2.7 Transient stability results for two-mass, three-mass and six-mass models (considering all types of damping, Condition 3)

IG POWER (MW)	1LG fault			2LS fault			2LG fault			3LG fault		
	2M	3M	6M	2M	3M	6M	2M	3M	6M	2M	3M	6M
50	O	O	O	O	O	O	O	O	O	O	O	O

2.3.4.4 Comparison Using Model System II

Other simulation results using model system II are shown here, in which longer fault clearing times and circuit breaker reclosing times are used. A fault occurs at 0.1 sec, the circuit breakers (CB) on the faulted line are opened at 0.22 sec and at 1.22 sec, are reclosed. Table 2.3 shows the initial values used in the simulation. A 3LG fault is considered to occur at fault point F of Fig. 2.17. The responses of the IG rotor and the turbine hub speeds for Conditions 4 and 5 are shown in Figs. 2.36 and 2.37, respectively. It is seen that all types of drive train models show similar characteristics for both Condition 4 and 5. Therefore, it is clear that the three drive train models show similar transient characteristics for both stable and unstable conditions of a WTGS. Some other simulation results are presented in Figs. 2.38 and 2.39 for a 2LG fault using Condition 4, in which the transformer near the induction generator is grounded or ungrounded, respectively. In these results, the six, three, and two mass drive train models also show the same transient characteristics during a network disturbance.

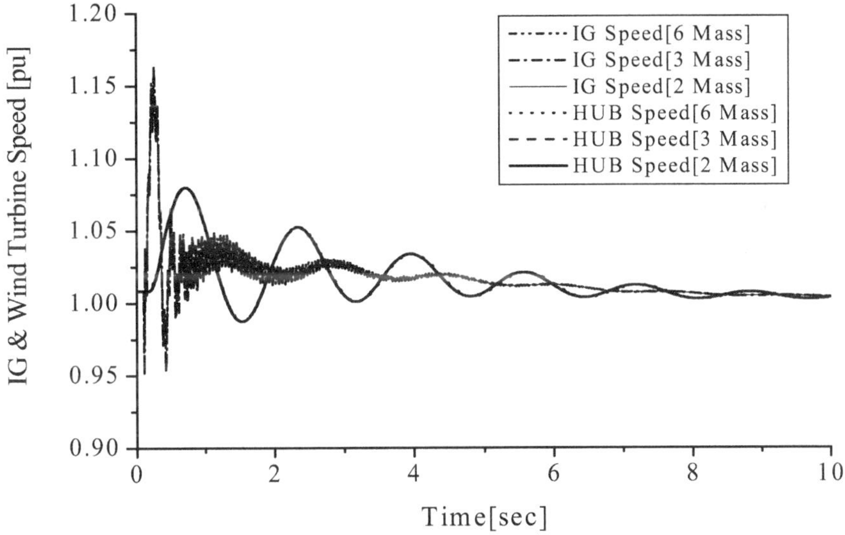

Fig. 2.36 Speed responses of the six, three, and two-mass models (Condition 4, 3LG, model system II)

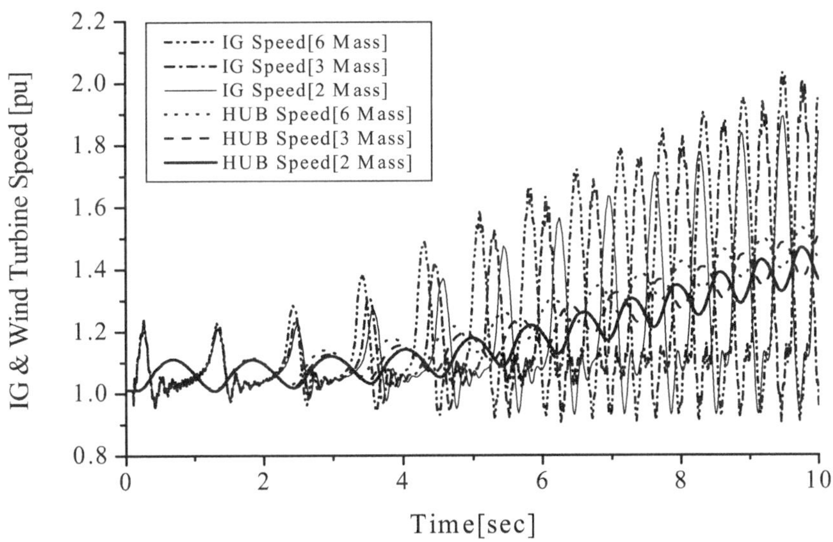

Fig. 2.37 Speed responses of the six, three, and two-mass models (Condition 5, 3LG, model system II)

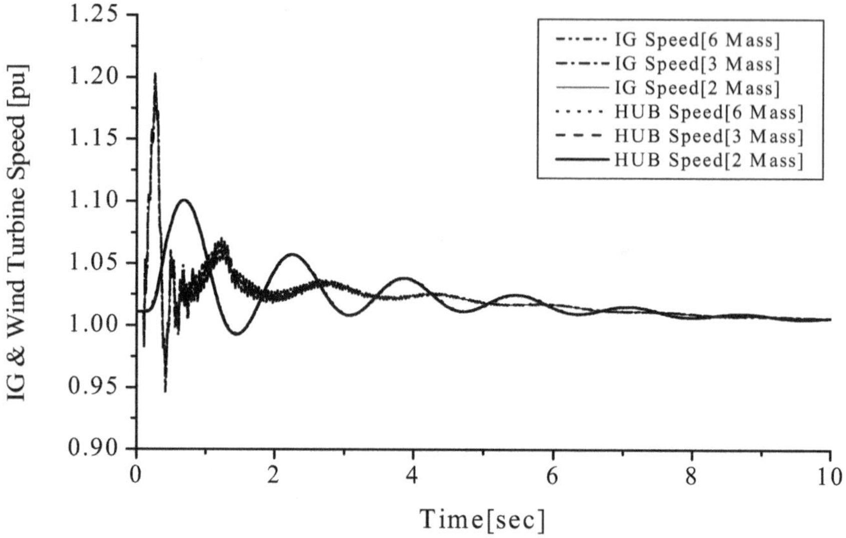

Fig. 2.38 Speed responses of the six, three, and two-mass models (Condition 4, 2LG, model system II, transformer grounded)

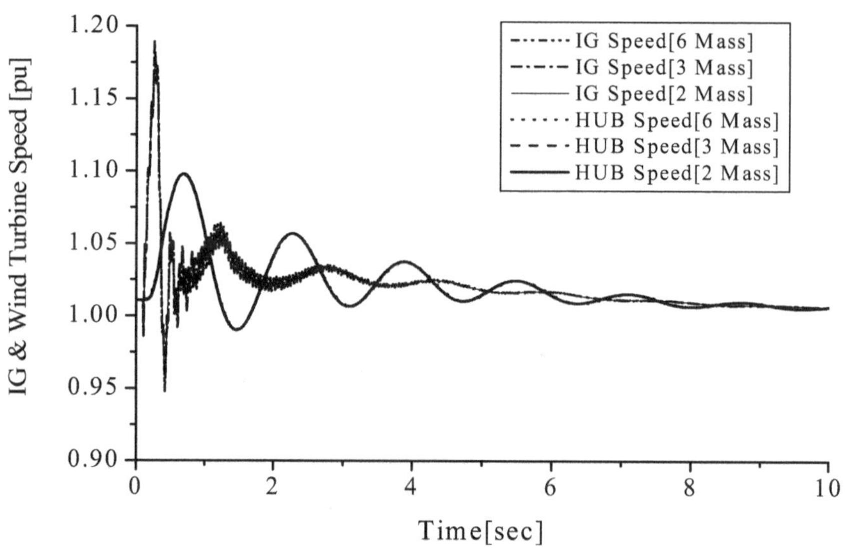

Fig. 2.39 Speed responses of the six, three, and two-mass models (Condition 4, 2LG, model system II, transformer ungrounded)

2.4 Variable Speed WTGS

Another commercial trend of a wind power generation is in using variable speed wind turbine (VSWT) driving a doubly fed induction generator (DFIG), wound field synchronous generator (WFSG) or permanent magnet synchronous generator (PMSG). The main advantage of variable speed operation is that more energy can be generated for a specific wind speed regime. Although the electrical efficiency decreases due to the losses in the power electronic converters that are essential for variable speed operation, the aerodynamic efficiency increases due to variable speed operation [119]. The aerodynamic efficiency gain can exceed the electrical efficiency loss, resulting in a higher overall efficiency [127, 128]. In addition, the mechanical stress is less because the rotor acts as a flywheel (storing energy temporarily as a buffer), reducing the drive train torque variations. Noise problems are reduced as well because the turbine runs at low speed. The main drawback of variable speed generating systems is that they are more expensive. However, using a variable speed generating system can also give major savings in other subsystems of the turbine such as lighter foundations in offshore applications, limiting the overall cost increase.

2.4.1 Variable Speed Topological Overview

The currently available variable speed wind turbine generator system topologies are shown in Fig. 2.40. To allow variable speed operation, the mechanical rotor speed and the electrical frequency of the grid must be decoupled. Therefore, a power electronic converter is used in a variable speed wind turbine generator system. In the doubly fed induction generator, a back-to-back voltage source converter feeds the three-phase rotor winding. In this way, the mechanical and electrical rotor frequency are decoupled, and the electrical stator and rotor frequencies can be matched independently of the mechanical rotor speed. In the direct drive synchronous generator system (PMSG or WFSG), the generator is completely decoupled from the grid by a frequency converter. The grid side of this converter is a voltage source converter, i.e., an IGBT (insulated gate bipolar transistor) bridge. The generator side can be either a voltage source converter or a diode rectifier.

The generator is excited using either an excitation winding (in the case of a WFSG) or permanent magnets (in the case of PMSG). In addition to these three mainstream generating systems, there are some other varieties, as explained in [119, 120].

One that must be mentioned here is the semi-variable speed system. In a semi-variable speed turbine, a winding type induction generator of which the rotor resistance can be changed by power electronics is used. By changing the rotor

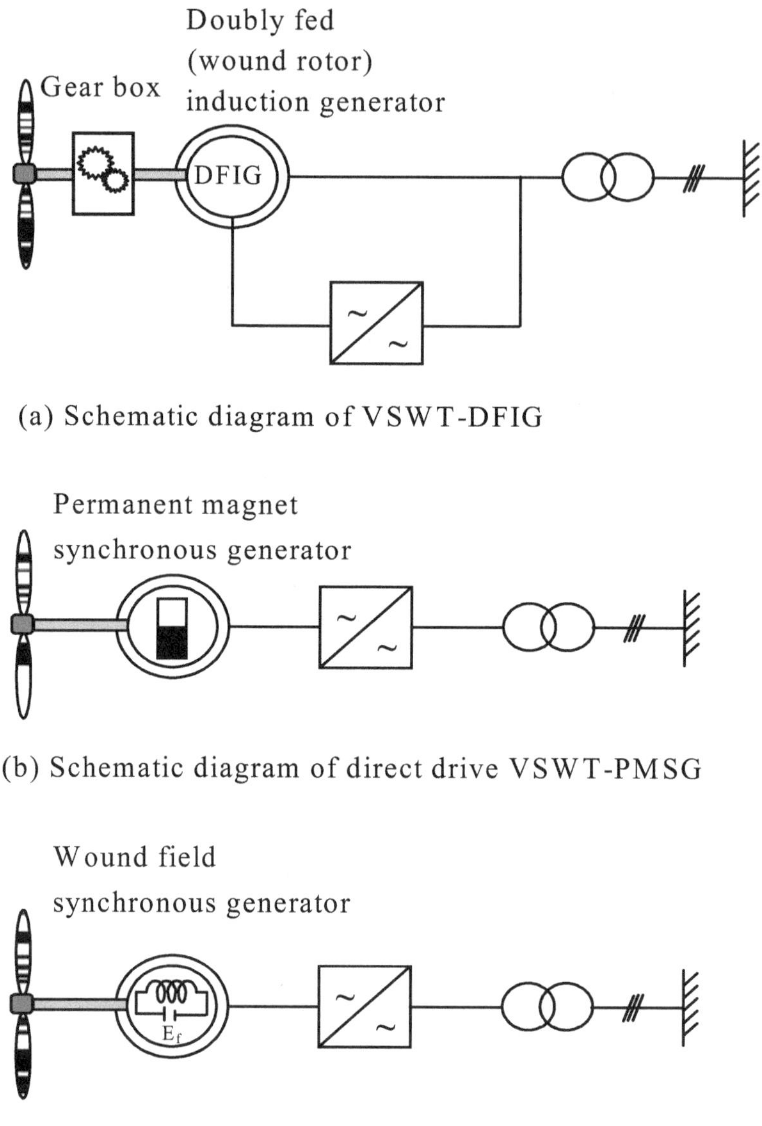

Fig. 2.40 Commercially available variable speed wind turbine generator systems

resistance, the torque/speed characteristic of the generator is shifted, and about a 10 % rotor speed decrease from the nominal rotor speed is possible. In this generating system, a limited variable speed capability is achieved at relatively low cost.

Other variations are a squirrel cage induction generator and a conventional synchronous generator connected to the wind turbine through a gearbox and to the grid by a power electronics converter of the full generator rating.

2.4.2 Variable Speed Wind Turbine Characteristics

To calculate Cp for the given values of β and λ, the following numerical approximations [34, 120] have been used in this study.

$$\lambda = \frac{\omega_r R}{V_w} \tag{2.28a}$$

$$\lambda_i = \frac{1}{\dfrac{1}{\lambda + 0.02\beta} - \dfrac{0.03}{\beta^3 + 1}} \tag{2.28b}$$

$$C_p(\lambda, \beta) = 0.73\left[\frac{151}{\lambda_i} - 0.58\beta - 0.002\beta^{2.14} - 13.2\right] e^{\frac{-18.4}{\lambda_i}} \tag{2.28c}$$

For a VSWT, generated active power depends on the power coefficient, C_p, which is related to the proportion of power extracted from the wind hitting the wind turbine blades. From Eq. 2.28c, the optimum values of tip speed ratio and power coefficient are chosen as 5.9 and 0.44 respectively. For each instantaneous wind speed of a VSWT, there is a specific turbine rotational speed, that corresponds to the maximum active power from the wind generator. In this way, the maximum power point tracking (MPPT) for each wind speed, increases the energy generation in a VSWT.

In this book, the 2.5 MW wind turbine with a rotor diameter of 84 m is considered. Its power coefficient curve with MPPT is shown in Fig. 2.41, from which it can be seen that, for any particular wind speed, there is a rotational speed, ω_r, that corresponds to the maximum power, P_{max}.

When the wind speed changes, the rotational speed is controlled to follow the maximum power point trajectory. Note here that precise measurement of wind speed is difficult. Therefore, it is better to calculate the maximum power, P_{max}, without measuring wind speed, as shown below:

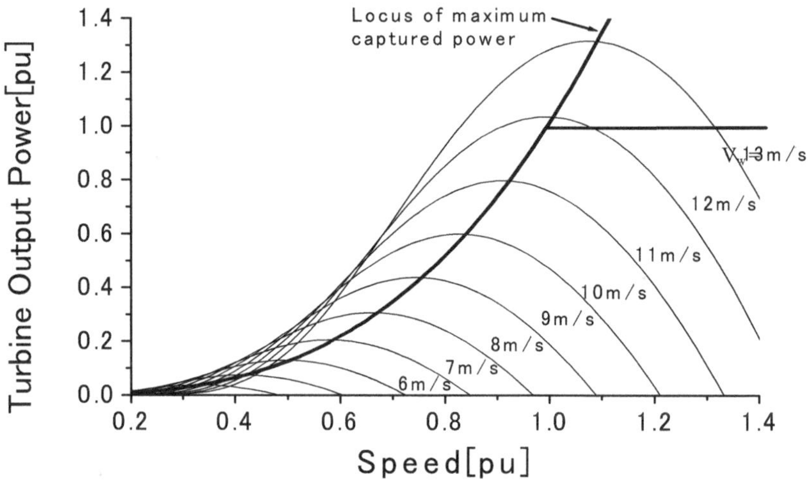

Fig. 2.41 Turbine characteristic with maximum power point tracking

$$P_{max} = 0.5\rho\pi R^2 \left(\frac{\omega_r R}{\lambda_{opt}} \right)^3 C_{p\_opt} \tag{2.29}$$

From Eq. 2.29, it is clear that the maximum power generated is proportional to the cube of the rotational speed as shown below:

$$P_{max} \propto \omega_r^3 \tag{2.30}$$

2.4.3 Influence of Drive Train Modeling on Variable Speed WTGS

For a fixed speed WTGS, detailed drive train dynamics might be considered, especially in transient analysis, as mentioned in Sect. 2.3.3. In a VSWT, however, the drive train properties have almost no effect on the grid side characteristics due to the decoupling effect of the power electronic converter [34]. Therefore, in the analyses of variable speed wind turbine generator system, the simple one-mass lumped model is considered in this book.

2.5 Chapter Summary

In this chapter, the basic theory of mechanical power extraction from wind is described briefly. Then fixed and variable speed wind turbine systems are explained in detail. The initial value calculation method for a power system including a WTGS, is also described.

Then emphasis is given to the drive train modeling of a fixed-speed WTGS. Three different types of drive train models are presented, and a comparative study is carried out among those models. A detailed transformation methodology from the six-mass to two-mass drive train models is presented, which can be used in simulation analysis with reasonable accuracy. By using the transformation procedure the inertia constants, spring constants, the self damping of individual masses and mutual damping of adjacent masses of the six-mass drive train model can be converted to reduced order models. The effects of drive train parameters, such as inertia constants, spring constants, and damping constants are examined for the above mentioned three-types of drive train models. It is proved that the two-mass drive train model of a WTGS is sufficient enough for the transient stability analysis of a fixed speed WTGS. The commercially available fixed and variable speed WTGS topologies are also shown in this chapter.

Chapter 3

Pitch Controller

To investigate the impacts of the integration of fixed speed wind farms into utility networks, transient stability should be analyzed before connecting a wind turbine generator system (WTGS) to the power system. In this chapter, a new logical pitch controller equipped with a fuzzy logic controller (FLC) has been proposed that can enhance the transient performance of a WTGS during severe network disturbances. Moreover, it can maintain the output power at the rated level when the wind speed is higher than the rated speed. To evaluate the effectiveness of the proposed controller in improving the transient stability, simulations have been carried out for severe network disturbances and severe wind conditions, considering the mechanical dead zone of the pitch actuation system.

The wind generator has an undesirable characteristic that its output power fluctuates randomly due to wind speed variation. This fluctuation can be decreased significantly by changing the blade pitch angle of the wind turbine. In this chapter, another new pitch controller based on fuzzy logic control is proposed that can smooth the wind generator's output power fluctuation. The wind generator's output power loss and smoothness level are analyzed when the proposed pitch controller is used in a wind turbine system. Comparative studies are carried out using three types of input command power in the controller. Moreover, different types of wind speed patterns are used to validate the effectiveness of the proposed controller. Simulation results show that the wind power fluctuation can be reduced well by using the proposed fuzzy logic based pitch controller.

This chapter has three main sections as follows:

- Conventional pitch controller.
- Fuzzy logic controlled pitch controller with power and speed control mode.
- Wind generator's power smoothing by using the new pitch controller.

3.1 Conventional Pitch Controller

The conventional pitch controller shown in Fig. 3.1 can be used to maintain the output power of a wind generator at its rated level when the wind speed is over the rated speed. In some studies, this pitch controller is used to enhance the transient stability of a WTGS when a network disturbance occurs in the power system.

The pitch servo is modeled with a first order delay system with a time constant, T_d. Because the pitch actuation system cannot, in general, respond instantly, a rate limiter is added to obtain a realistic response. The limitations of this pitch controller are described in Chap. 1 of this book.

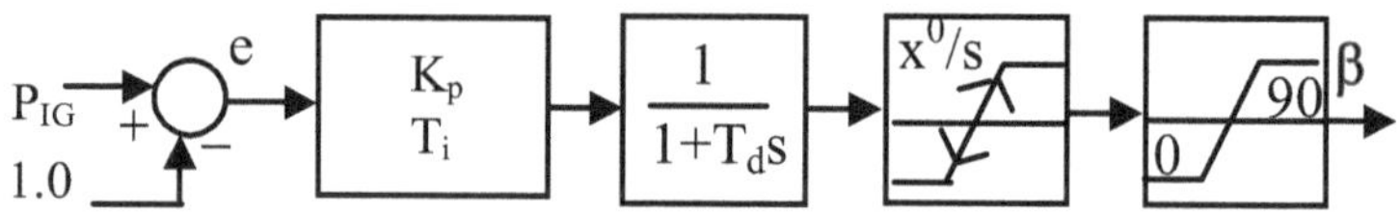

Fig. 3.1 Conventional pitch controller

3.2 Fuzzy Logic Controlled Pitch Controller with Power and Speed Control Modes

The main purpose of using a pitch controller with a wind turbine is to maintain a constant output power at the terminal of the wind generator (in this case, induction generator, IG, is considered as wind generator) when the wind speed is higher than the rated speed. The proposed controller shown in Fig. 3.2 can serve this purpose well. Moreover, it can enhance the transient stability of an induction generator. The controller input is normally set to INPUT1 and it works in the power control mode, where P_{IG}^{REF} is a reference value for the generator output and is varied according to the terminal voltage of an induction generator because the induction generator cannot generate rated power when its terminal voltage is below the rated voltage. When the terminal voltage is sensed as a controller input, a low pass filter might be necessary to reduce harmonics of terminal voltage. The transfer function of the low pass filter, $F_{LP}(s)$, is shown in Eq. 3.1 where the values of gain, G, damping ratio, ζ, and characteristic frequency, f_c ($\omega_c = 2\pi f_c$), are chosen as 1.0, 0.7, and 60.0 Hz respectively.

$$F_{LP}(s) = \frac{G}{1 + 2\zeta(s/\omega_c) + (s/\omega_c)^2} \tag{3.1}$$

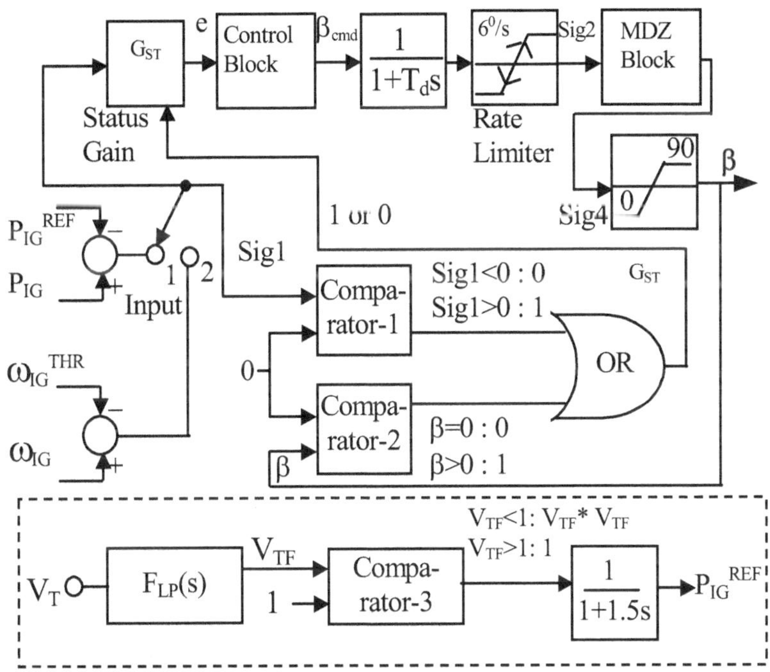

Fig. 3.2 Fuzzy logic controlled logical pitch controller

Table 3.1 Status gain determination

	Sig1< 0	**Sig1> 0**
β=0	0	1
β>0	1	1

On the other hand, if the IG rotor speed increases to a threshold value, i.e., 3 % increase from its rated speed, the controller input will be set to INPUT2 and it works in the speed control mode, where ω_R^{THR} is the threshold value. The operating status of the pitch controller will be determined by a status gain, G_{ST}, which is the output of the logical comparator. The construction of the logical comparator is very simple, as shown in Fig. 3.2. The output of the logical comparator can be determined from Table 3.1.

The pitch control system can be electric or hydraulic, individual or global pitch [39]. The pitch servo is modeled with a first order system [13, 14, 20, 34, 35, 38, 40] with a time constant, T_d. Because recently the servomotor can operate very fast, a servo system delay of 0.25 sec and 0.2 sec is chosen, respectively, in [20, 38]. But probably there might be some other delays, i.e., communication delay, computational delay, and conditional delay (to overcome Coulomb friction) that might take a few hundred milliseconds more. That's why in this work T_d is chosen as 1.5 sec, which is sufficient to consider all types of delays in the pitch actuation system. It is also important to mention that the pitch actuation system cannot respond instantly. The pitch rate commanded by the actuator is physically limited to ± 10°/s at the maximum [14, 20, 34, 35, 37, 38, 41, 42]. In [20], a pitch rate of ± 5°/s is considered, but the transient performance of the pitch controller is not analyzed there. In the speed control mode, the larger pitch rate value shows better transient performance. In this work, the rate limiter value of ± 6°/s has been chosen to obtain a realistic response.

Another feature that makes the proposed controller more practical is the inclusion of a mechanical dead zone (MDZ) block in the pitch actuation system of Fig. 3.2, which is shown in detail in Fig. 3.3. To reduce actuator motion for a longer lifetime and to eliminate noise in the command signal, the dead zone is necessary to be considered when the commanded pitch rate is less than ± 0.1°/s. The MDZ block is designed in such a way that it will pass or hold the rate limiter output depending on whether the pitch rate is above or below 0.1°/s, respectively, as shown in Fig. 3.3. In the previous works [10, 14, 19 – 21, 33 – 43, 120], the MDZ block is not considered in the modeling of pitch controller. Moreover, the power and speed control modes are not shown separately and the terminal voltage of wind generator is not sensed as the controller input. The logic circuit unit is also not shown in those works.

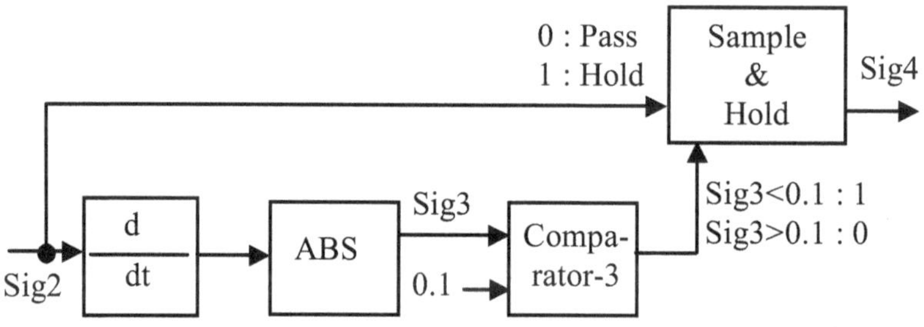

Fig. 3.3 Modeling of the mechanical dead zone

3.2.1 Controller Design Phase

As a control methodology of the proposed pitch controller, FLC and PI controllers are investigated. Simulation results show that a FLC gives better performance than a PI controller in all operating conditions. It is known that a fuzzy controller can work well with a non-linear system [129, 130]. Because the wind turbine characteristics are quite non-linear and wind speed is intermittent and stochastic by nature, we propose a pitch controller equipped with a FLC.

3.2.1.1. Fuzzy Logic Controller Design

The proposed FLC system shown in Fig. 3.4 is used to find the angle, β_{cmd}, in the control block in Fig. 3.2 from the error signal, e, and the change of error signal, Δe. The FLC is explained in the following.

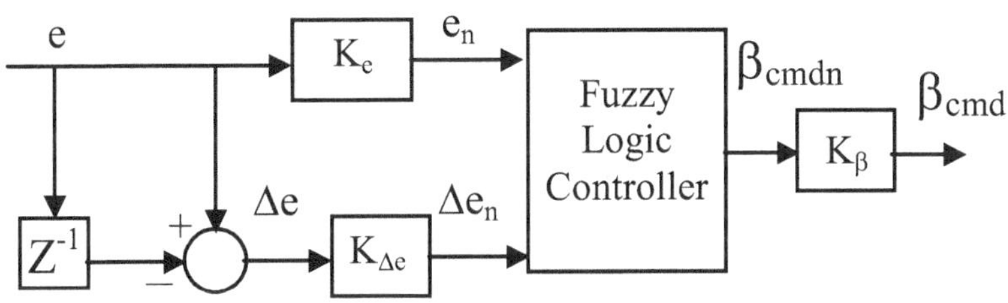

Fig. 3.4 Structure of a fuzzy logic controller

3.2.1.1.1 Fuzzification

To design the proposed FLC, the error signal, e(k), and the change of error signal, $\Delta e(k)$ are considered the controller inputs. The angle, β_{cmd}, is considered the controller output, which is actually the pitch angle command signal for the mechanical servo system. For convenience, the inputs and output of the FLC are scaled with coefficients K_e, $K_{\Delta e}$, and K_β, respectively. These scaling factors can be constants or variables and play an important role in the FLC design to achieve a good response in both transient and steady states. In this work, these scaling factors are considered constant for simplicity of the controller design and are selected by trial and error. The values of K_e, $K_{\Delta e}$, and K_β are chosen as 1.0, 1000, and 100, respectively.

In Fig. 3.4, Z^{-1} represents one sampling time delay. The triangular membership functions with overlap used for the input and output fuzzy sets are shown in Fig. 3.5 in which the linguistic variables are represented by NB (Negative Big), NS (Negative Small), Z (Zero), PS (Positive Small), and PB (Positive Big). The grade of input membership functions can be obtained from the following equation [129]:

$$\mu(x) = [w - 2|x - m|]/w \tag{3.2}$$

where, $\mu\ (x)$ is the value of the grade of membership, w is the width, m is the co-ordinate of the point at which the grade of membership is 1, and x is the value of the input variable.

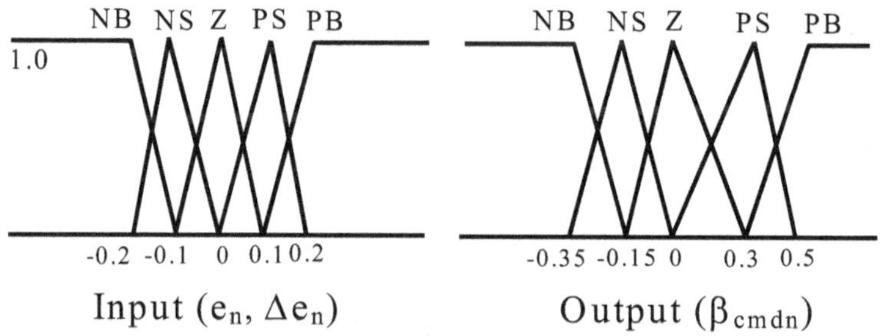

Fig. 3.5 Fuzzy sets and their corresponding membership functions

3.2.1.1.2 Rule Base

The fuzzy mapping of the input variables to the output is represented by IF-THEN rules of the following forms:

IF < e_n is NB> and <Δe_n is NB> THEN < β_{cmdn} is NB>.
IF < e_n is ZO> and <Δe_n is ZO> THEN < β_{cmdn} is ZO>.
IF < e_n is PB> and <Δe_n is PB> THEN < β_{cmdn} is PB>.

The entire rule base is given in Table 3.2. There is a total of 25 rules to achieve the desired angle, β_{cmd}.

Table 3.2 Fuzzy rule table

β_{cmdn}		Δe_n				
		NB	NS	ZO	PS	PB
e_n	NB	NB	NB	NS	NS	ZO
	NS	NB	NS	NS	ZO	PS
	ZO	NS	NS	ZO	PS	PS
	PS	NS	ZO	PS	PS	PB
	PB	ZO	PS	PS	PB	PB

3.2.1.1.3 Inference and Defuzzification

In this work, Mamdani's max-min (or sum-product) [129] method is used for the inference mechanism. The center of gravity method [129] is used for defuzzification to obtain β_{cmdn}, which is given by the following equation:

$$\beta_{cmdn} = \sum_{i=1}^{N} \mu_i C_i \Big/ \sum_{i=1}^{N} \mu_i \tag{3.3}$$

where, N is the total number of rules, μ_i is the membership grade for the i-th rule and C_i is the coordinate corresponding to the respective output or consequent membership function [$C_i \in \{-0.35, -0.15. 0.0, 0.3, 0.5\}$]. The actual modulated angle, β_{cmd}, can be found by multiplying β_{cmdn} by the scaling factor K_β.

3.2.1.2 PI Controller Design

The classical PI controller finds extensive application in industrial control. The structure of a continuous time PI controller used as the control block in Fig. 3.2 is shown in Fig. 3.6, where e (the error signal, i.e., power or speed) is the input and β_{cmd} is the output of the PI controller. K_P and T_i represent the proportional gain and integration time constant respectively. The values of K_P and T_i chosen are 100.0 and 0.3, respectively.

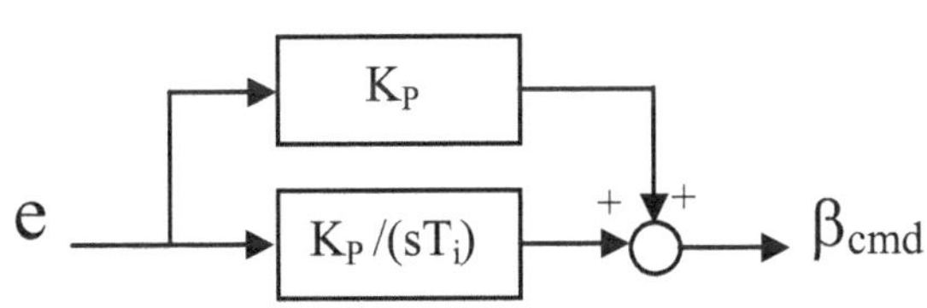

Fig. 3.6 Structure of a PI controller

3.2.2 Model System Used in Sect. 3.2

Figure 3.7 shows the model system used for the simulation of the transient stability analysis of a WTGS. Here, one synchronous generator (SG) is connected to an infinite bus through a transformer and a double circuit transmission line. In the figure, the double circuit transmission line parameters are numerically shown in the form of R+jX, where R and X represent the resistance and reactance, respectively. One wind farm (Induction generator, IG) is connected to the network via a transformer and a short transmission line. A single cage induction generator is considered in this analysis to obtain the worst-case scenario. A capacitor bank has been used for reactive power compensation at steady state. The value of capacitor C is chosen so that the power factor of the wind power station becomes unity when it is operating in the rated condition (V=1.0, P=0.5) [43]. The AVR (automatic voltage regulator) and GOV (governor) control system models shown in Figs. 2.13 and 2.14, respectively (Sect. 2.3.4.1 of Chap. 2) are used in the synchronous generator model in the simulation. Generator parameters are shown in Table 3.3. The system base is 100 MVA. The initial values used in the simulation are shown in Table 3.4. Condition 1 and Condition 2 were obtained by the Case I method, and Condition 3 was obtained by the Case II method explained in Sect. 2.3.1.2 of Chap. 2. The fixed speed wind turbine characteristics are described in Chap. 2.

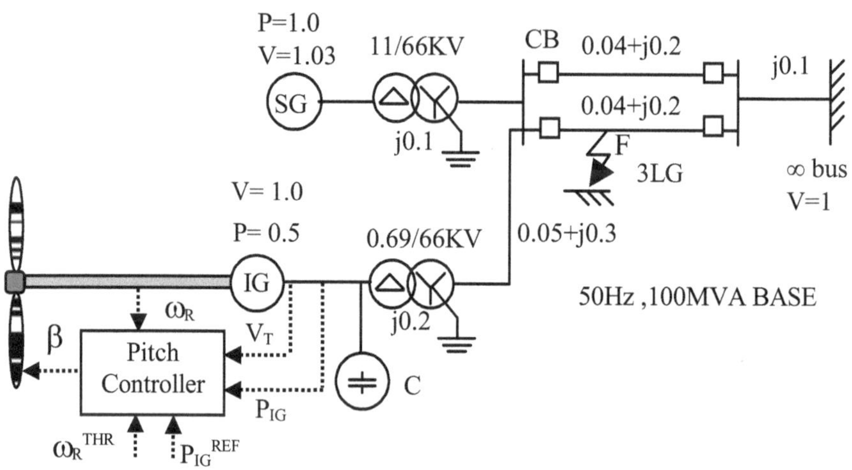

Fig. 3.7 Model system

3.2.3 Simulation Results for Sect. 3.2

Four cases have been considered for performance analysis of the proposed pitch controller. A time step of 0.00005 sec has been chosen, and simulation time has been chosen as 300 sec for Case 1A & Case 3, 15 sec for Case 1B, and 50 sec for Case 2. The initial values used in the simulations for Cases 1A and 1B have been taken from Condition 1 of Table 3.4, and the initial values for Case 2 and Case 3 have been taken from Condition 2 and Condition 3, respectively, of the same table. For transient performance analysis a 3LG fault is considered to occur at point F of Fig. 3.7. The simulations were done by using PSCAD/EMTDC[1] [126].

Table 3.3 Generator Parameters

SG		IG	
MVA	100	MVA	50
ra (pu)	0.003	r1 (pu)	0.01
xa (pu)	0.13	x1 (pu)	0.1
Xd (pu)	1.2	Xmu (pu)	3.5
Xq (pu)	0.7	r2 (pu)	0.01
X_d' (pu)	0.3	x2 (pu)	0.12
X_q' (pu)	0.22	H(sec)	1.5
X_d'' (pu)	0.22		
X_q'' (pu)	0.25		
T_{do}' (sec)	5.0		
T_{do}'' (sec)	0.04		
T_{qo}'' (sec)	0.05		
H (sec)	2.5		

Table 3.4 Initial conditions of generators and turbines

	Condition 1		**Condition 2**		**Condition 3**	
	SG	**IG**	**SG**	**IG**	**SG**	**IG**
P(pu)	1.0	0.285	1.0	0.50	1.0	0.50
V(pu)	1.03	1.08	1.03	0.992	1.03	0.999
Q(pu)	0.170	0.111 (0.196)*	0.384	0.004 (0.264)*	0.334	0.00 (0.263)*
Efd(pu)	1.652	-	1.851	-	1.803	-
Tm(pu)	1.002	-	1.003	-	1.003	-
δ (deg)	50.17	-	59.11	-	50.71	-
slip	0.0	−0.523%	0.0	−1.13%	0.0	−1.11%
Vw (m/s)	-	9.46	-	11.80	-	13.20
β (deg)	-	0	-	0	-	9.77

* Reactive power drawn by induction generator

[1] For the latest information on PSCAD/EMTDC, visit at http://pscad.com

3.2.3.1 Case 1A

The objective of this case is to demonstrate the power and speed control modes of the proposed controller at low wind speed as shown in Fig. 3.8, which are the real wind speed data obtained on Hokkaido Island, Japan. A 3LG fault of 0.1 sec duration is considered to occur at 50 sec when the wind speed is less than the rated speed. Responses of real power, terminal voltage, rotor speed of induction generator and blade pitch angle are shown in Figs. 3.9 – 3.12, respectively. It is seen that the IG speed doesn't exceed the threshold value after the disturbance, and it becomes stable for both cases with and without the pitch controller.

When the wind speed increases above its rated speed, then the IG without a pitch controller cannot maintain the output power at the rated level, as shown in Fig. 3.9. But the IG with the proposed controller can maintain the output power at the rated level. The pitch controller equipped with a FLC can work well in the power control mode with a lower overshoot compared to the pitch controller equipped with a PI controller. This will be clear in Sect. 3.2.3.4.

Fig. 3.8 Wind speed (Cases 1A and 1B)

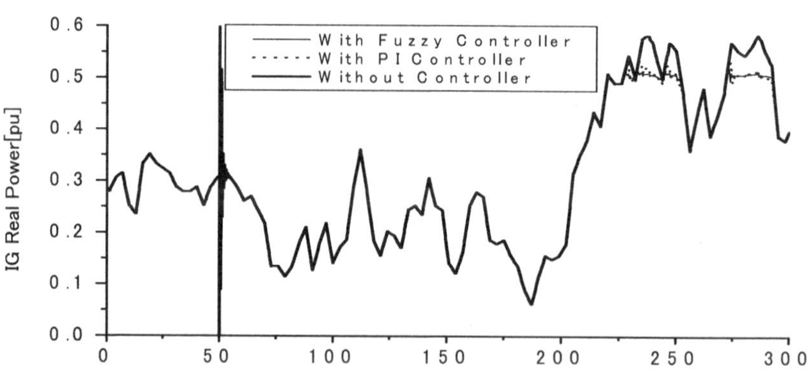

Fig. 3.9 Real power of the induction generator (Case 1A)

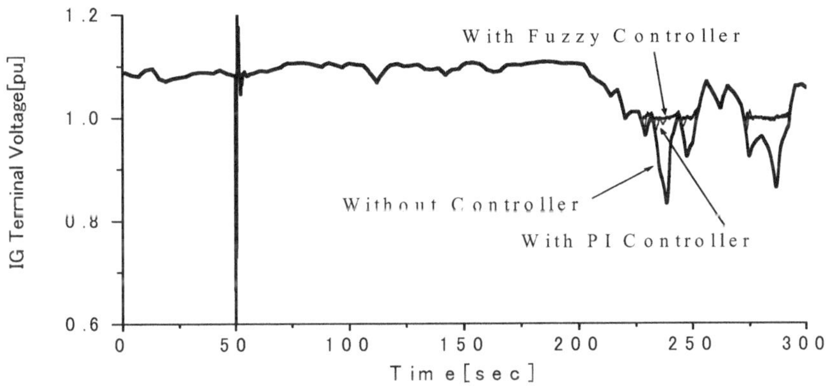

Fig. 3.10 Terminal voltage of the induction generator (Case 1A)

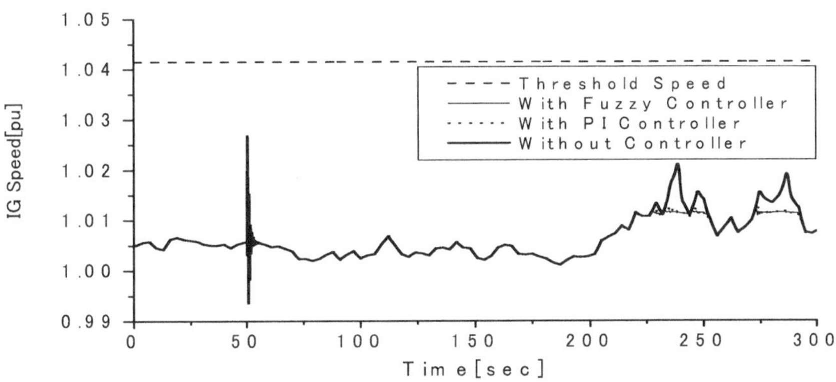

Fig. 3.11 Rotor speed of the induction generator (Case 1A)

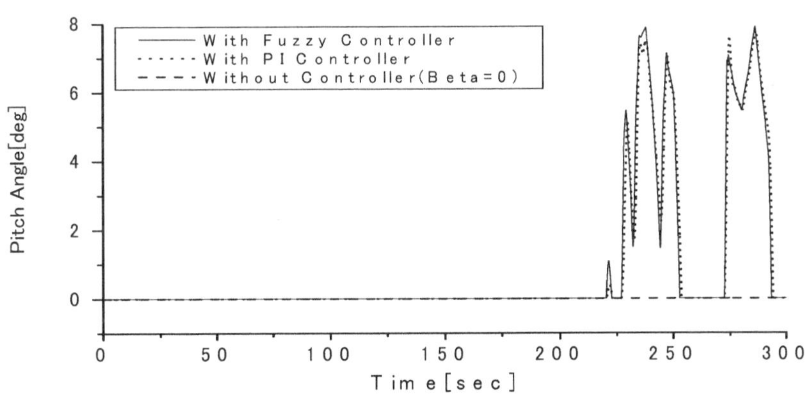

Fig. 3.12 Pitch angle of the wind turbine (Case 1A)

3.2.3.2 Case 1B

In this case, the transient performance of the induction generator with the proposed pitch controller is analyzed. The fault occurs at 220.1 sec in Fig. 3.8, when the wind speed is at the rated level. The circuit breakers (CB) on the faulted line are opened at 220.2 sec and are re-closed at 221.0 sec. After the fault occurs, the IG rotor speed starts to increase rapidly, as shown in Fig. 3.13. When the rotor speed exceeds the threshold value, then the pitch controller works in the speed control mode, and the IG becomes stable again. But without a controller the IG goes out of step. The IG real power and terminal voltage with and without a controller are shown in Figs. 3.14 and 3.15, respectively. The wind turbine pitch angle and load angle of synchronous generator are shown in Figs. 3.16 and 3.17, respectively. It is noticeable that the synchronous generator doesn't go out of step when the induction generator is unstable. In this case, the FLC gives a better response than a conventional PI controller from the viewpoint of settling time.

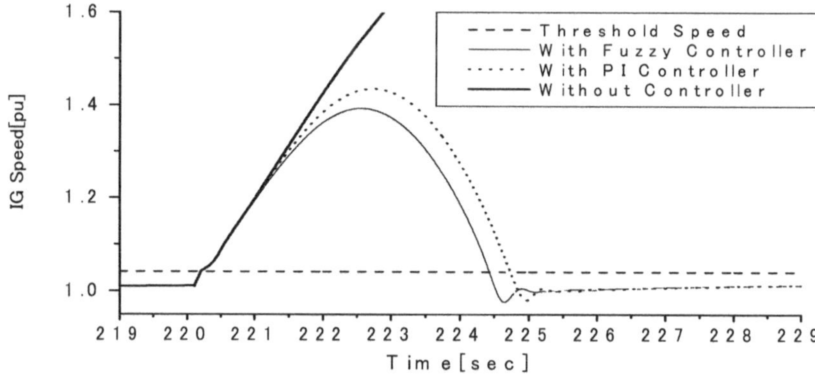

Fig. 3.13 Rotor speed of the induction generator (Case 1B)

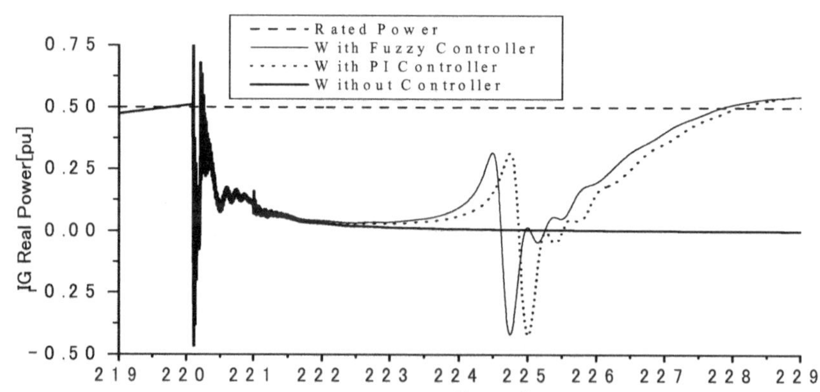

Fig. 3.14 Real power of the induction generator (Case 1B)

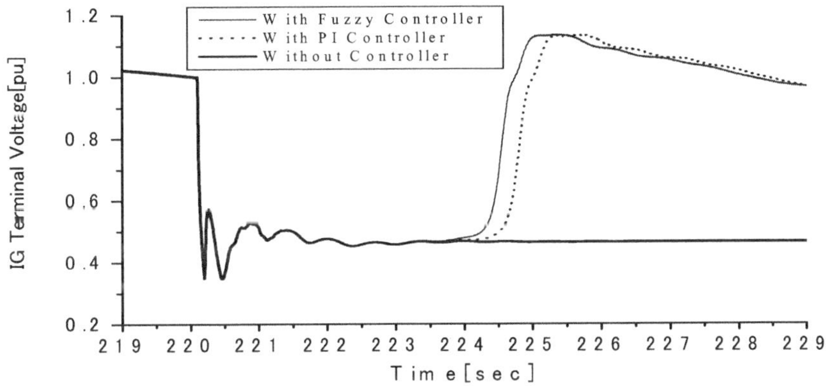

Fig. 3.15 Terminal voltage of the induction generator (Case 1B)

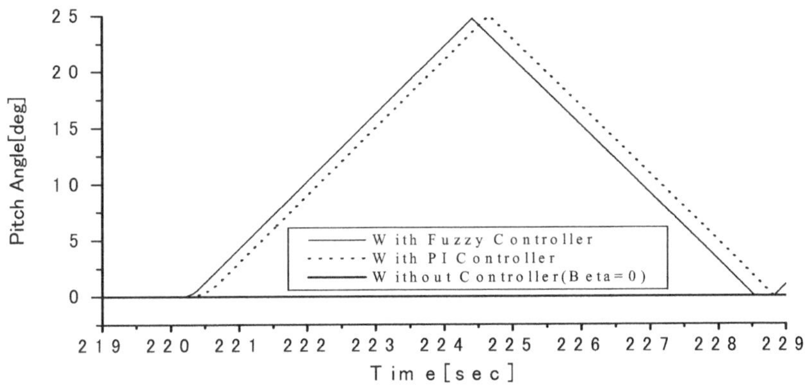

Fig. 3.16 Blade pitch angle (Case 1B)

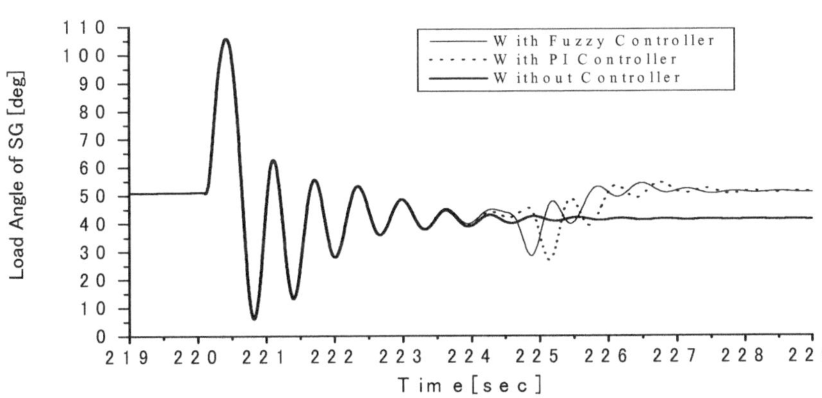

Fig. 3.17 Load angle of the synchronous generator (Case 1B)

3.2.3.3 Case 2

In this case, the necessity of taking the terminal voltage of the induction generator as the pitch controller input is demonstrated. Depending on the network parameters or fault conditions, there can be some situations in which the terminal voltage of a wind generator should be taken as the pitch controller input. For example, we consider the case where the double circuit transmission line parameters in Fig. 3.7 are just doubled. So, the power transfer capability of the network will be decreased, and the network disturbance will be more severe. The circuit breakers at both ends of one line are considered to be opened at 0.1 sec and remain open for a long time, 40 sec. Then the circuit breakers are closed again. Real wind speed data shown in Fig. 3.18 are used here. This case is analyzed in three different ways: (1) no controller is used; (2) the proposed pitch controller is used without a voltage sensing unit (shown in Fig. 3.19a), i.e., the reference power is always remaining constant at the rated power; and (3) the proposed controller is used with a voltage sensing unit (shown in Fig. 3.19b), where the reference power varies according to the terminal voltage of the induction generator. The responses of the terminal voltage and rotor speed of the induction generator are shown in Figs. 3.20 and 3.21, respectively. The response of the turbine blade pitch angle is shown in Fig. 3.22. The IG without the pitch controller becomes unstable.

When the pitch controller without the terminal voltage sensing unit is used, the IG rotor cannot become stable because at low terminal voltage the IG cannot generate the rated power. Therefore, it is necessary to change the reference power of the pitch controller according to the terminal voltage of the IG. This has been clearly presented in Figs. 3.19 – 3.22. Moreover, the proposed pitch controller with a FLC unit can make the IG stable more quickly than the pitch controller with a PI unit as shown in Fig. 3.21.

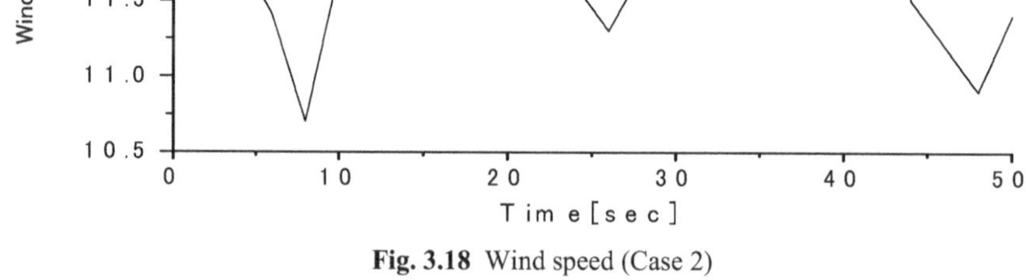

Fig. 3.18 Wind speed (Case 2)

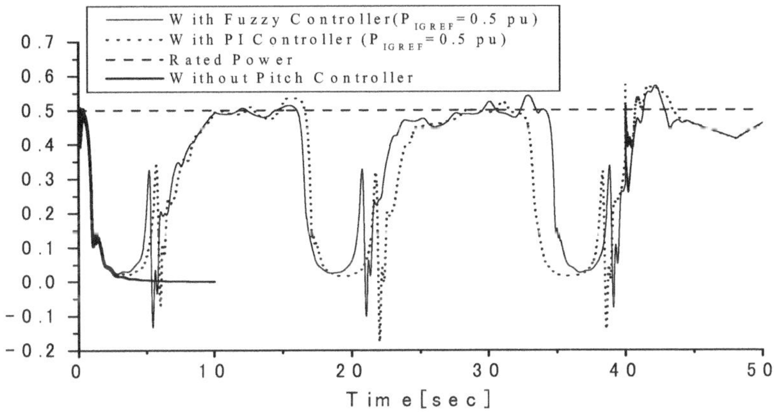

Fig. 3.19(a) Real power of the IG without a voltage sensing unit (Case 2)

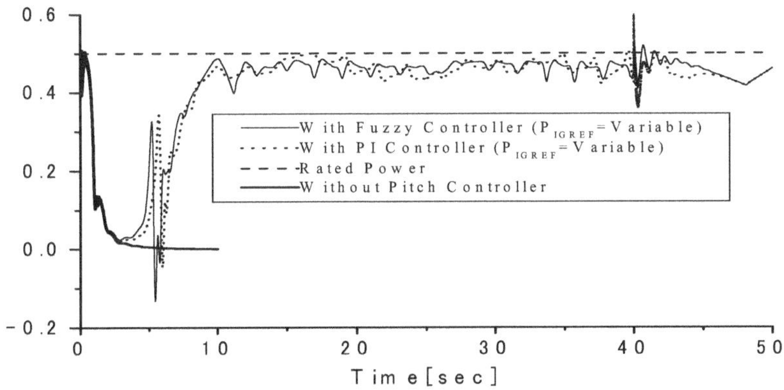

Fig. 3.19(b) Real power of the IG with a voltage sensing unit (Case 2)

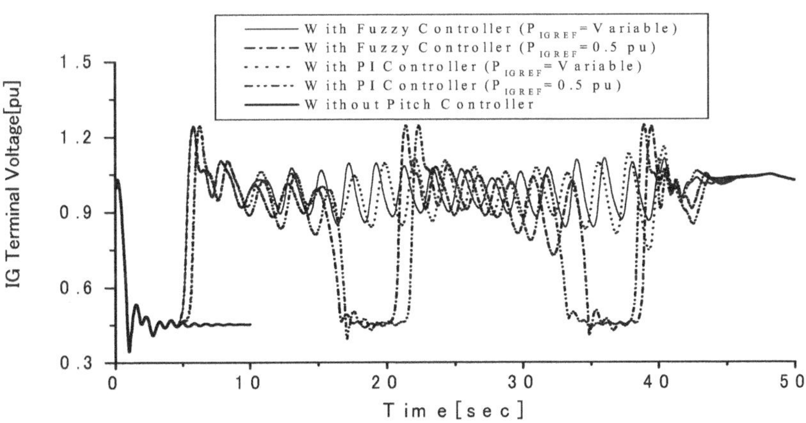

Fig. 3.20 Terminal voltage of the induction generator (Case 2)

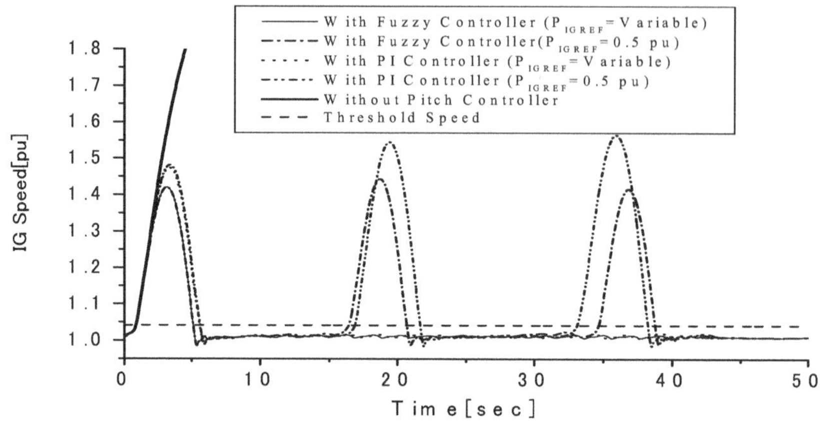

Fig. 3.21 Rotor speed of the induction generator (Case 2)

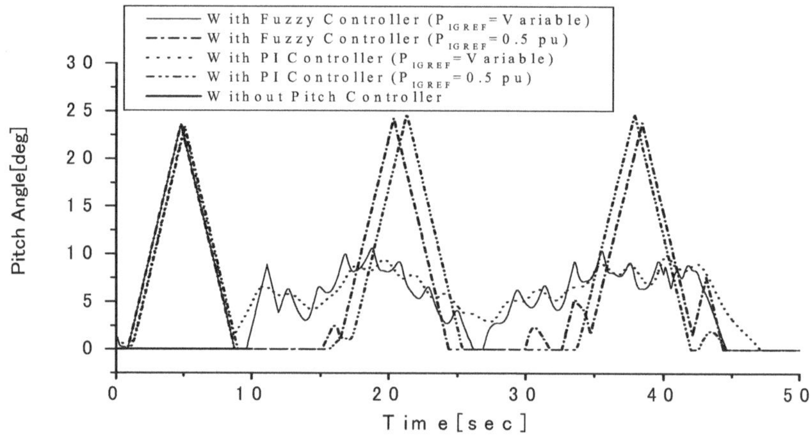

Fig. 3.22 Blade pitch angle (Case 2)

3.2.3.4 Case 3

In this case, pitch controller performance is evaluated by using another wind speed pattern shown in Fig. 3.23, where the wind speed is fluctuating more frequently than those in Fig. 3.8 or Fig. 3.18. It is noticeable that the initial wind speed is 13.2 m/s, which is above the rated speed shown in Table 3.4.

To evaluate the transient performance of the proposed pitch controller, a 3LG fault is considered to occur at point F in Fig. 3.7. The fault occurs at 150.0 sec, the circuit breakers (CB) on the faulted line are opened at 150.1 sec, and are closed at 151.0 sec. The responses of real power, terminal voltage, rotor speed of the IG, and the blade pitch angle of the wind turbine are shown in Figs. 3.24 – 3.27, respectively.

It is seen that the IG without the pitch controller cannot maintain the output power at the rated level, and it goes out of step, though there is no network disturbance. In contrast, using the proposed controller the IG output power can be maintained at the rated level when the wind speed is above the rated speed.

When a 3LG fault of 0.1 sec duration occurs at 150 sec, the pitch controller enters the speed control mode. The IG terminal voltage can return to its pre-fault value and becomes stable. The pitch controller equipped with a FLC unit can make the IG stable more quickly compared to that with a PI unit.

It is noticeable that at 216 sec the when wind speed rapidly increases, the FLC equipped pitch controller can control the output power without switching to the speed control mode. On the other hand, the PI equipped pitch controller enters the speed control mode at this severe condition to make the IG stable as the rotor speed goes above the threshold value. Moreover, the pitch controller equipped with a FLC unit can also reduce the power and voltage fluctuations significantly compared to that with a PI unit, as shown in Figs. 3.24 and 3.25, respectively.

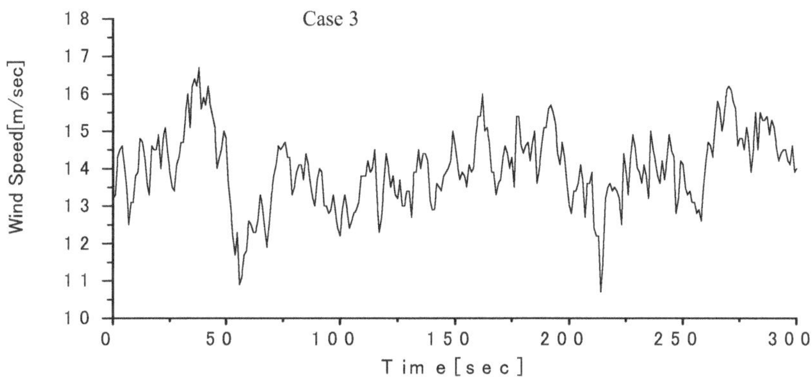

Fig. 3.23 Wind speed (Case 3)

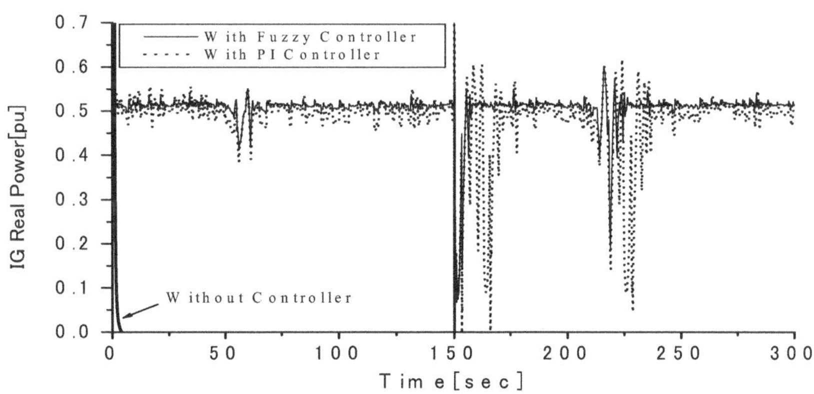

Fig. 3.24 Real power of the induction generator (Case 3)

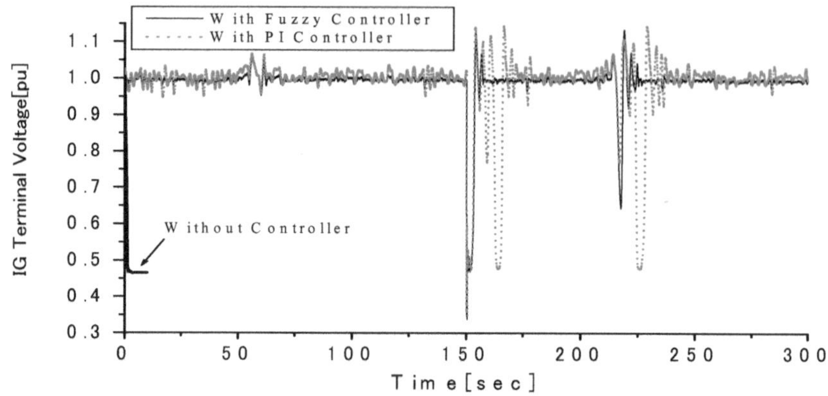

Fig. 3.25 Terminal voltage of the induction generator (Case 3)

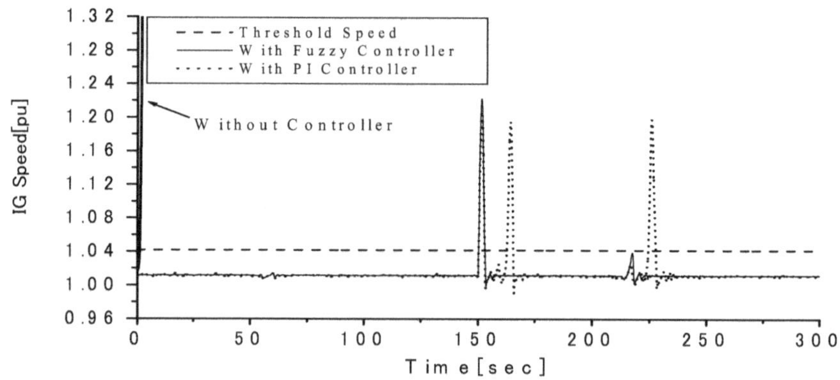

Fig. 3.26 Rotor speed of the induction generator (Case 3)

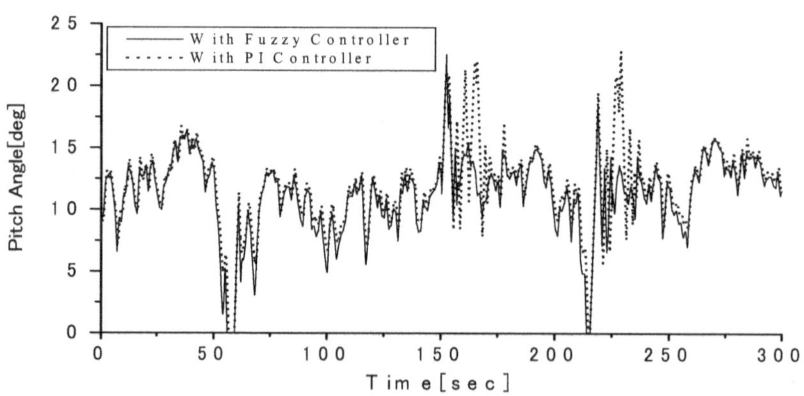

Fig. 3.27 Blade pitch angle (Case 3)

3.3 Wind Generator Power Smoothing by Using the New Pitch Controller

3.3.1 Calculating Controller Input Power Command, P_{IG}^{REF}

For wind generator output power smoothing, the most important part is to determine the pitch controller input power command, P_{IG}^{REF}. The turbine characteristic described in Chap. 2 is necessary for calculating the input power command. Three types of average values are evaluated in this work to ensure the effectiveness of the proposed controller.

3.3.1.1 Average (AVG)

This value is calculated after every specified number of periods. For twenty measurements from M_1 through M_{20}, the successive four period average values, for example, are as follows:

$$
\begin{aligned}
AVG_4 &= (M_4 + M_3 + M_2 + M_1)/4 \\
AVG_8 &= (M_8 + M_7 + M_6 + M_5)/4 \\
&\cdot \\
&\cdot \\
AVG_{20} &= (M_{20} + M_{19} + M_{18} + M_{17})/4
\end{aligned}
\tag{3.4a}
$$

3.3.1.2 Simple Moving Average (SMA)

The n period simple moving average for period number d is computed from

$$
SMA_d = \frac{\sum_{i=1}^{n} M_{(d-i)+1}}{n} \quad (n \le d) \tag{3.4b1}
$$

If ten measurements, M_1 through M_{10}, are available, then the successive four period simple moving averages, for example, are as follows:

$$
\begin{aligned}
SMA_4 &= (M_4 + M_3 + M_2 + M_1)/4 \\
SMA_5 &= (M_5 + M_4 + M_3 + M_2)/4 \\
&. \\
&. \\
SMA_{10} &= (M_{10} + M_9 + M_8 + M_7)/4
\end{aligned} \tag{3.4b2}
$$

It is not possible to compute a four period moving average until four periods of data are available. That's why the first moving average in the above example is SMA_4.

3.3.1.3 Exponential Moving Average (EMA)

The formula for an exponential moving average is

$$
EMA(C) = [(C - P) \times K] + P \tag{3.4c}
$$

where,

C=The current value,
P=The previous period's EMA, and
K=Weighting factor.

For a period-based EMA, "K" is equal to 2/(1 + N), where N is the specified number of periods. For example, a 10-period EMA "weighting factor" is calculated like this: 2/(1+10)=0.1818.

The above-mentioned average values are demonstrated in Fig. 3.28. Sixty periods (180 sec) AVG, SMA, and EMA of wind speed are shown there. SMA starts from 180 sec when 60 periods of data are available. For the very first period EMA calculation, SMA is used. It is seen that because AVG is constant every 180 sec, it cannot follow a rapid wind speed change. On the other hand, the EMA can follow the wind speed trend more rapidly than the SMA because the EMA uses its previously calculated EMA value for the next calculation.

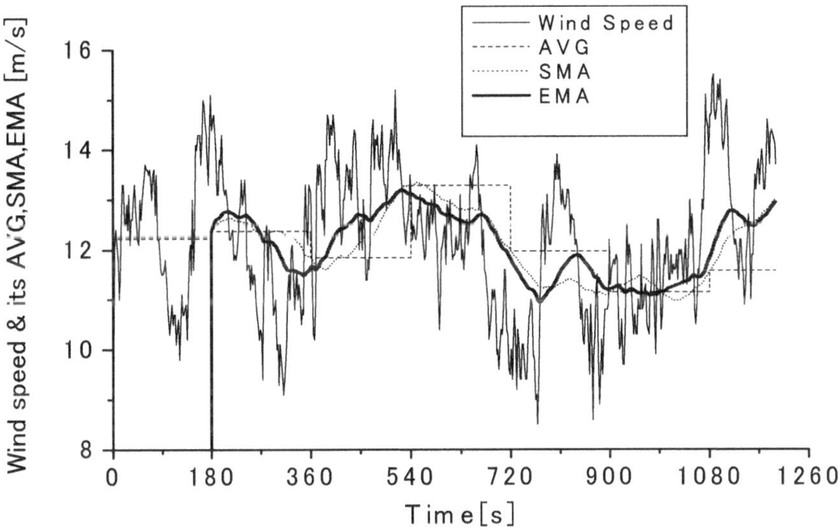

Fig. 3.28 Comparison among AVG, SMA, and EMA

The following steps explain the generation of the pitch controller input power command:

a. The wind turbine captured power, P_{WT}, can be obtained from Eq. 2.6.

b. The average value of wind turbine captured power, $\overline{P}_{WT}$, can be calculated from Eq. 3.4. In this paper, 60 periods average value with each period of 3 sec is used in the simulation, i.e., average time, T, of 180 sec is chosen.

c. The standard deviation can be calculated from the following equation:

$$P_{WT\sigma} = \sqrt{\frac{\int_{t-T}^{t} (P_{WT} - \overline{P}_{WT})^2 dt}{T}} \tag{3.5}$$

d. Finally, the controller's revised input power command, P_{IG}^{REF}, can be obtained from Eq. 3.6.

$$P_{IG}^{REF} = (\overline{P}_{WT} - P_{WT\sigma}) \tag{3.6}$$

The whole process is demonstrated in Fig. 3.29.

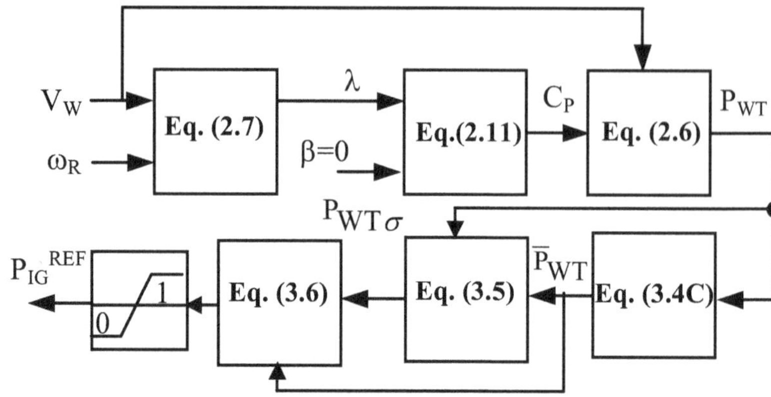

Fig. 3.29 Calculation of the controller input power command, P_{IG}^{REF}

3.3.2 Pitch Controller Design Phase

The wind turbine blade pitch angle is not controlled, in general, until the rated power is generated. When the wind speed is above the rated speed, then the pitch controller is activated to keep the output power at the rated level. In this section, a new pitch controller is presented where the turbine blade pitch angle is controlled even when the wind speed is below the rated speed. The proposed pitch controller is shown in Fig. 3.30. The pitch controller input power command, P_{IG}^{REF}, is generated from the average value of the wind turbine captured power, as explained before. Then the difference between P_{IG}^{REF} and P_{IG} is progressed through a fuzzy logic controller (FLC) to generate the command signal, β_{cmd}, for the mechanical servo system.

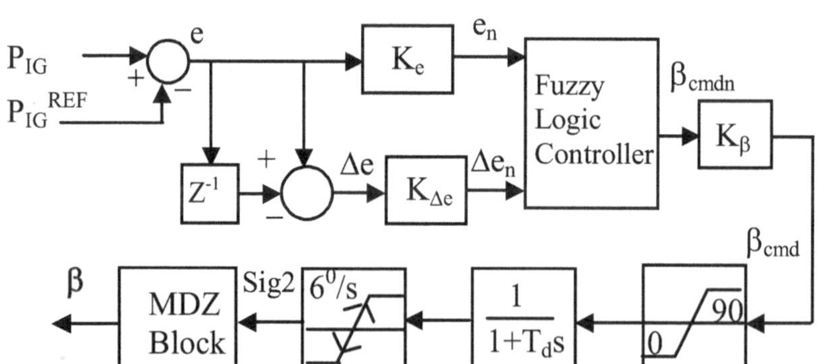

Fig. 3.30 Pitch controller for power smoothing

For wind power smoothing, the wind turbine blade needs to pitch frequently. Therefore, special care is needed in the design phase of the blade pitch actuation system. The servo system is designed as mentioned in Sect. 3.2. The rate limiter and mechanical dead zone are also considered as described in Sect. 3.2 for the sake of precise analysis.

As a control methodology of the proposed pitch controller, the FLC is adopted for wind power smoothing. Simulation results show that the FLC gives better performance in all operating conditions. The FLC is explained briefly in the next section.

For convenience, the inputs and output of the FLC are scaled with coefficients K_e, $K_{\Delta e}$, and K_β, respectively. The values of K_e, $K_{\Delta e}$, and K_β chosen are 1.0, 2000, and 285, respectively. The triangular membership functions with overlap used for the input and output fuzzy sets are shown in Fig. 3.31, in which the linguistic variables are represented by NB (Negative Big), NM (Negative Medium), NS (Negative Small), Z (Zero), PS (Positive Small), PM (Positive Medium), and PB (Positive Big). The grade of input membership functions can be obtained from Eq. 3.2 [129]. The entire rule base is given in Table 3.5. There is a total of 49 rules to

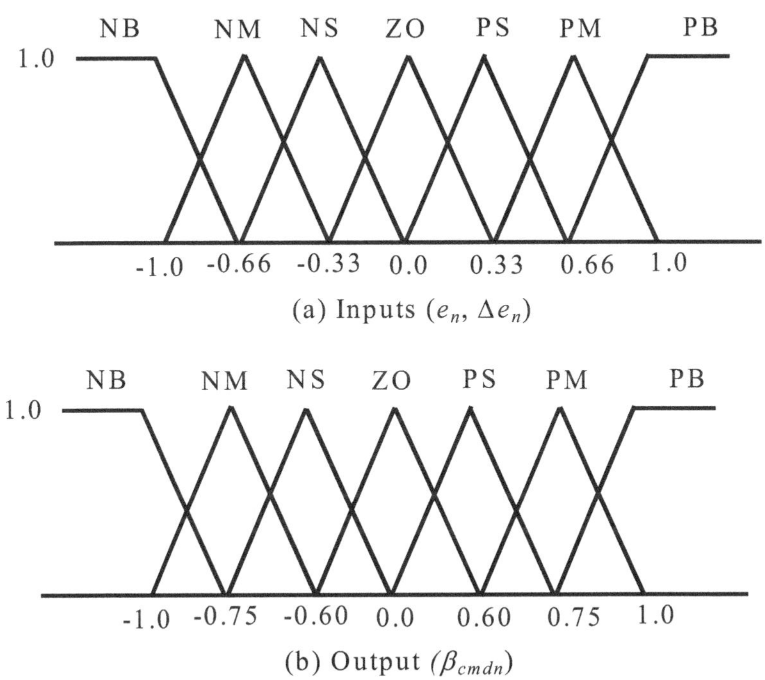

Fig. 3.31 Fuzzy sets and their corresponding membership functions

achieve the desired angle, β_{cmdn}. Mamdani's max-min (or sum-product) [129] method is used for the inference mechanism. The center of gravity method [129] is used for defuzzification to obtain β_{cmdn}, which is given by Eq. 3.3. In Eq. 3.3, C_i is the consequent membership function [$C_i \in \{-1.0, -0.75, -0.60, 0.0, 0.60, 0.75, 1.0\}$]. The angle, β_{cmd}, can be obtained by multiplying β_{cmdn} by the scaling factor K_β.

Table 3.5 Fuzzy rule table

β_{cmdn}		Δe_n						
		NB	NM	NS	ZO	PS	PM	PB
e_n	NB	NB	NB	NM	NM	NS	NS	ZO
	NM	NB	NM	NM	NS	NS	ZO	PS
	NS	NM	NM	NS	NS	ZO	PS	PS
	ZO	NM	NS	NS	ZO	PS	PS	PM
	PS	NS	NS	ZO	PS	PS	PM	PM
	PM	NS	ZO	PS	PS	PM	PM	PB
	PB	ZO	PS	PS	PM	PM	PB	PB

3.3.3 Energy Loss and Smoothing Estimation

Energy loss and smoothing performance of the proposed pitch controller are compared with those of the conventional pitch controller shown in Fig 3.1. Total energy generation, W, of the IG is evaluated from the following equation and energy loss can be calculated as a percentage with respect to that of the conventional pitch controller:

$$W = \int_0^t P_{IG}(t)dt \tag{3.7}$$

For smoothing level estimation, two methods are considered. One is the frequency spectrum of the wind generator output power, where the low magnitude indicates better smoothing. The second is the following equation that can be treated as an overall power-smoothing index.

$$P_{index} = \int_0^t \left| \frac{dP_{IG}(t)}{dt} \right| dt \tag{3.8}$$

where the difference in the induction generator output power between two adjacent sampling instants is added simultaneously throughout the simulation time.

Therefore, if Eq. 3.8 is applied to two different signals for an equal time span, then the low value indicates better smoothness because the smooth signal's accumulation would be small.

3.3.4 Model System Used in Sect. 3.3

The model system used in the simulation study for wind generator output power smoothing is shown in Fig. 3.32. The synchronous and induction generator parameters are the same as those used in Sect. 3.2.2. The AVR and GOV models shown in Sect. 2.3.4.1 of Chap. 2 are used in the synchronous generator model. The system base is 100 MVA.

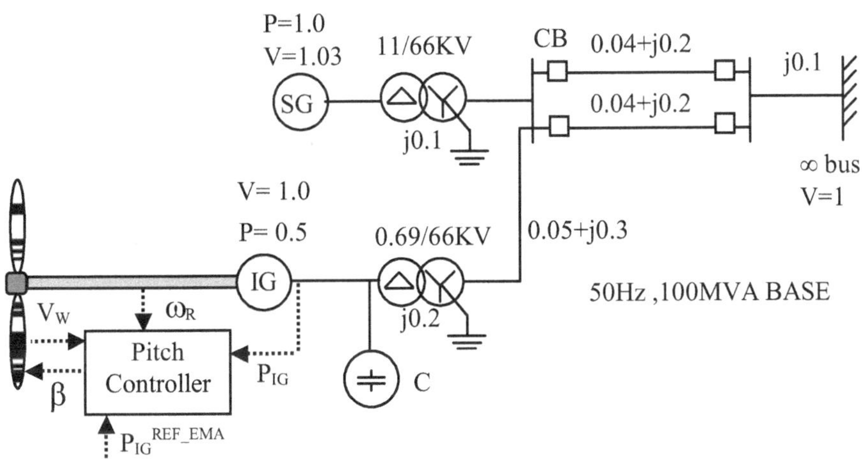

Fig. 3.32 Model system

3.3.5 Simulation Results for Sect. 3.3

A time step of 0.0001 sec and a simulation time of 600 sec have been chosen. In all the simulations, the pitch controller input power command is generated from 180 sec (60 periods, each of 3 sec) AVG, SMA, and EMA values that are expressed by $P_{IG}^{REF\_AVG}$, $P_{IG}^{REF\_SMA}$, and $P_{IG}^{REF\_EMA}$, respectively. For the first 180 sec (until 0 sec and not shown in the simulation results), $P_{IG}^{REF\_AVG}$ is used as the controller input power command when $P_{IG}^{REF\_SMA}$ and $P_{IG}^{REF\_EMA}$ are used. Therefore, simulations based on the three command signals can be performed from 0 sec. The simulation has been done by using PSCAD/EMTDC[2] [126]. To present the effectiveness of the proposed controller, the following cases are considered.

3.3.4.1 Case 1

In this case, the wind speed is always higher than the rated speed as shown in Fig. 3.33. The responses of the IG real power, the pitch controller input power command, and the blade pitch angle are presented in Figs. 3.34 – 3.36, respectively. Because the wind speed is always higher than the rated speed, three different input power commands of the proposed controller are the same. Therefore, only the results based on the EMA are presented. The FLC controlled pitch controller gives less oscillation compared to that of conventional pitch controller, which can be seen from the output power of the IG and its frequency spectrum shown in Figs. 3.34 and 3.37, respectively. The IG total energy generation obtained by using one of the controllers is presented in Fig. 3.38. Because the wind speed is always higher than the rated speed, almost the same energies are generated in both controllers.

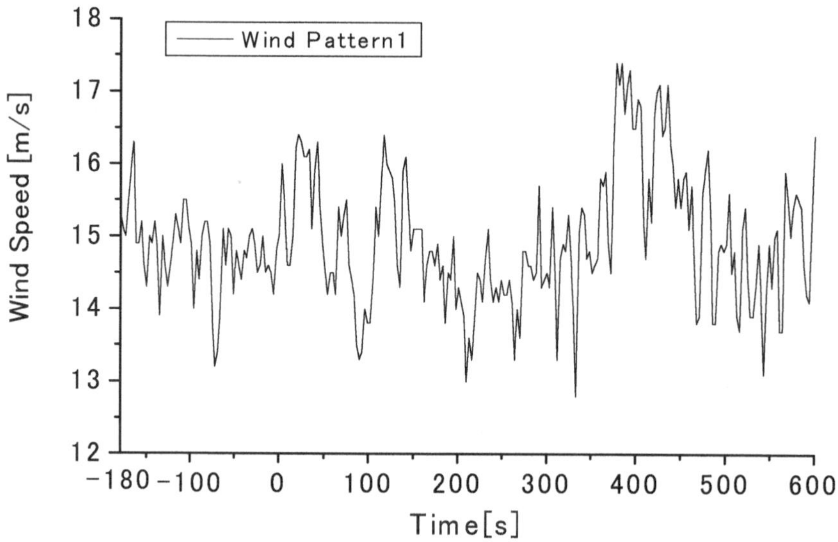

Fig. 3.33 Wind speed pattern 1 (Case 1)

---

[2] For the latest information on PSCAD/EMTDC, visit at http://pscad.com

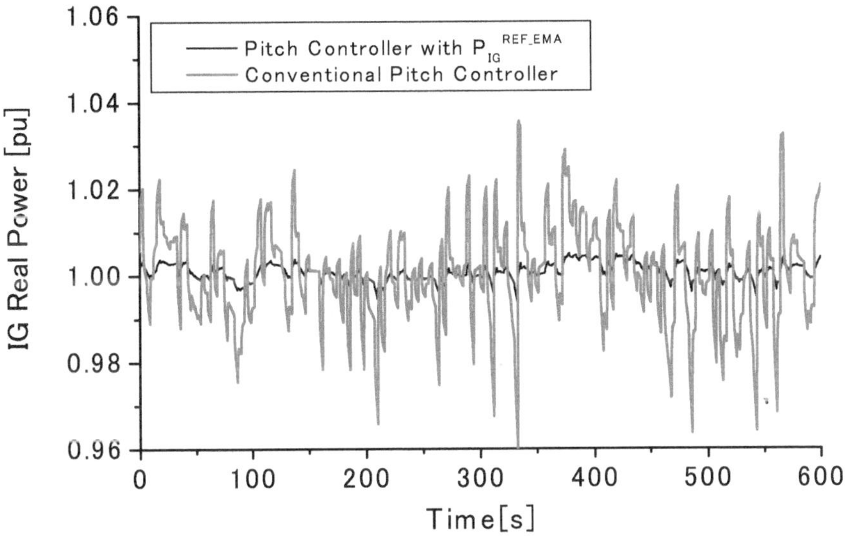

Fig. 3.34 Real power of the induction generator (Case 1)

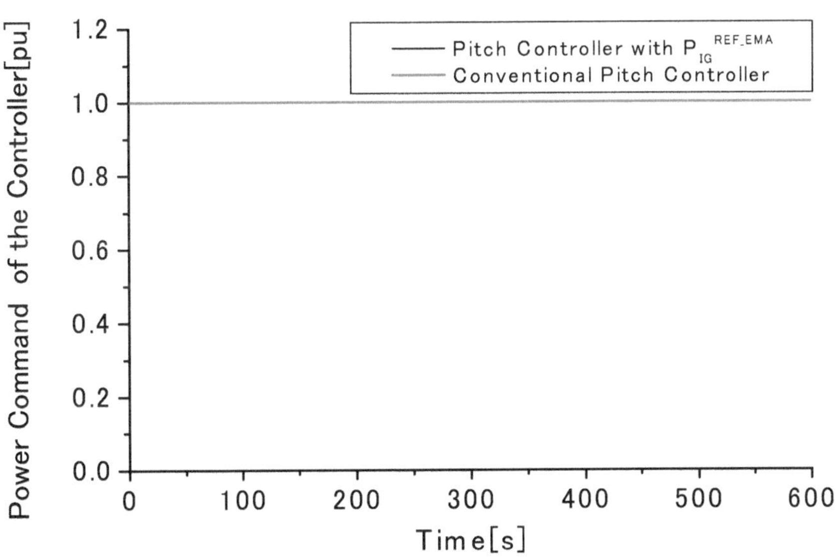

Fig. 3.35 Pitch controller input power command (Case 1)

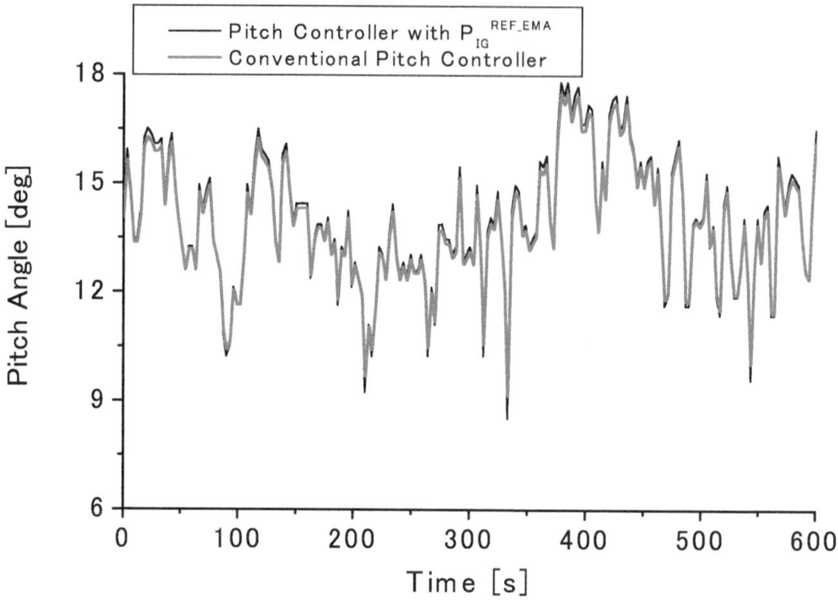

Fig. 3.36 Blade pitch angle of the wind turbine (Case 1)

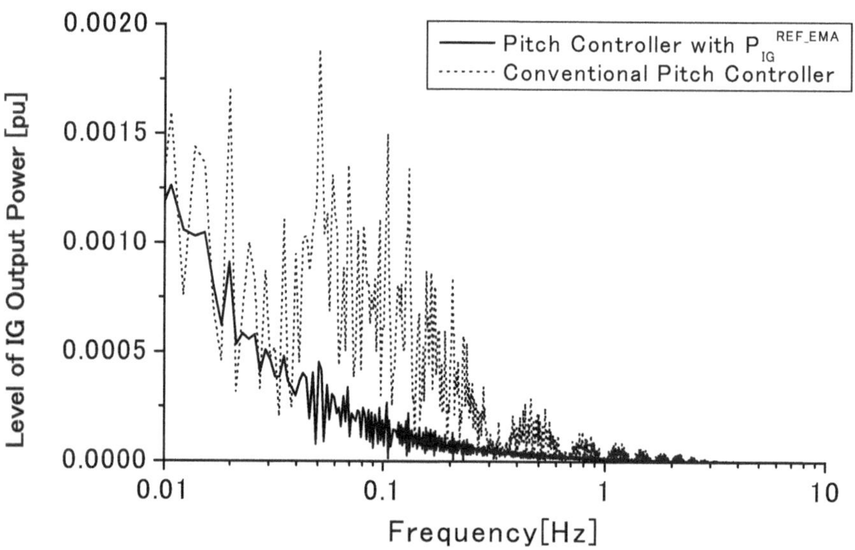

Fig. 3.37 Frequency spectrum of the IG output power (Case 1)

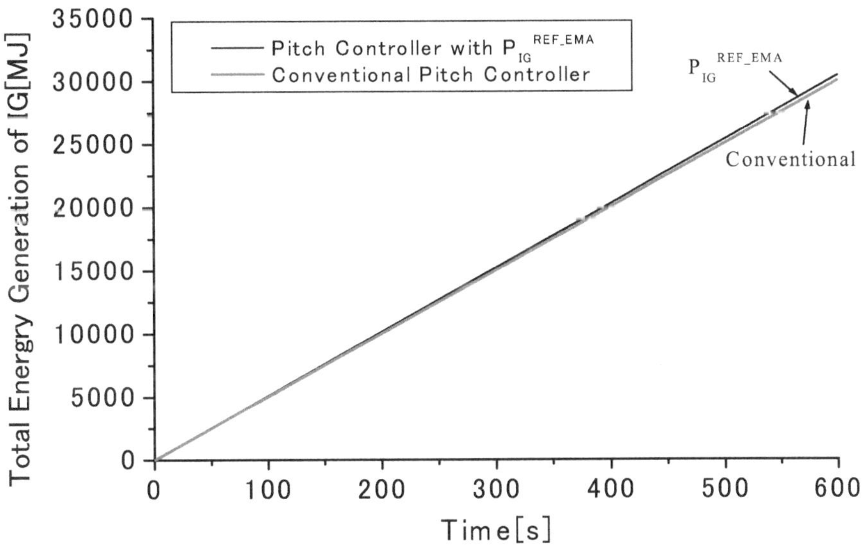

Fig. 3.38 Total energy generation by the induction generator (Case 1)

The mechanical dead zone has been considered in the simulations, as explained before. Table 3.6 shows the total mechanical dead time throughout the simulation time of 600 sec for three wind speed patterns. It is seen from Case 1 of Table 3.6 that for this wind pattern, the servo system stops the motion of the turbine blades for 65.20 sec to reduce the mechanical load on the turbine blades.

Table 3.6 Mechanical dead time

	AVG	**SMA**	**EMA**
Case 1	65.20	65.20	65.20
Case 2	18.01	17.02	17.24
Case 3	36.71	20.91	22.95

3.3.4.2 Case 2

In this case, the moderate wind speed pattern shown in Fig. 3.39 is used. The responses of the IG real power, the controller input power command, the blade pitch angle, and the frequency spectrum of the IG output are presented in Figs. 3.40 – 3.43, respectively. From the simulation results, it is clear that the FLC controlled pitch controller can smooth the wind generated power much better than the conventional pitch controller. The overall IG output smoothness function is also presented in Fig. 3.44 for pitch controller input power commands, $P_{IG}^{REF\_AVG}$, $P_{IG}^{REF\_SMA}$ and $P_{IG}^{REF\_EMA}$, where a lower value represents better smoothness. It is seen that using $P_{IG}^{REF\_AVG}$ and $P_{IG}^{REF\_EMA}$ as the pitch controller input power command give smoother results than $P_{IG}^{REF\_SMA}$.

The IG total energy generation for the conventional and proposed pitch controllers, obtained from Eq. 3.7 are presented in Fig. 3.45. In that figure, the percentage energy loss during 600 sec for each input power command is calculated with respect to the conventional pitch controller. The controller input power command of $P_{IG}^{REF\_EMA}$ gives the lowest energy loss among the three command signals.

The mechanical dead times of Case 2 for three different input power commands are shown in Table 3.6. They are less than those of wind pattern 1 because, in wind pattern 2, the wind speed takes a value both above and below the rated speed.

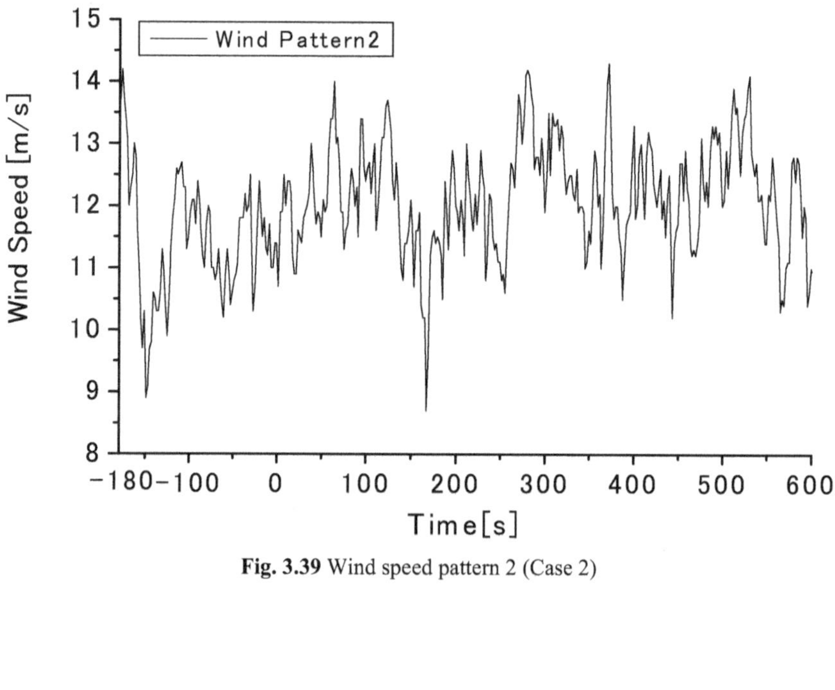

Fig. 3.39 Wind speed pattern 2 (Case 2)

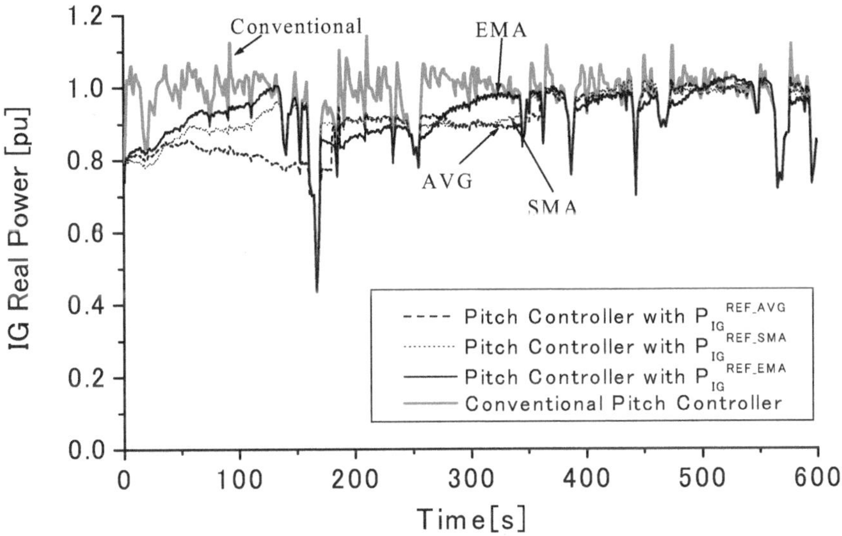

Fig. 3.40 Real power of the induction generator (Case 2)

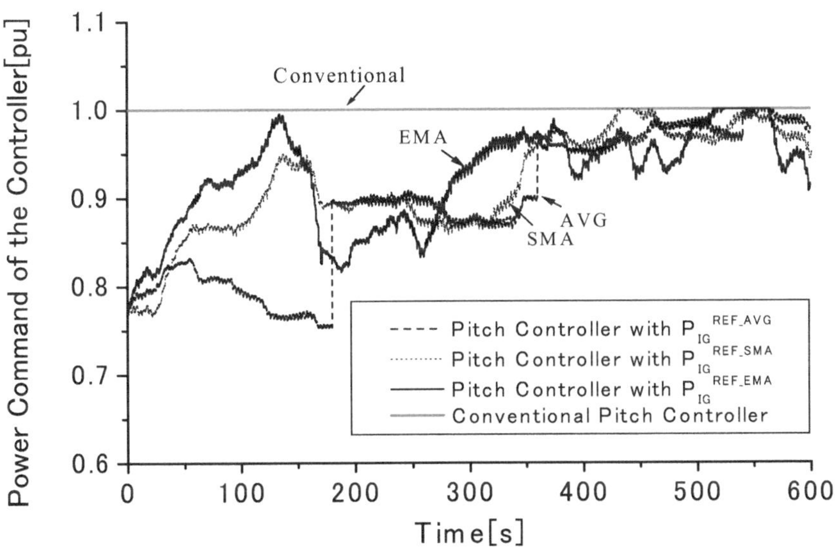

Fig. 3.41 Pitch controller input power command (Case2)

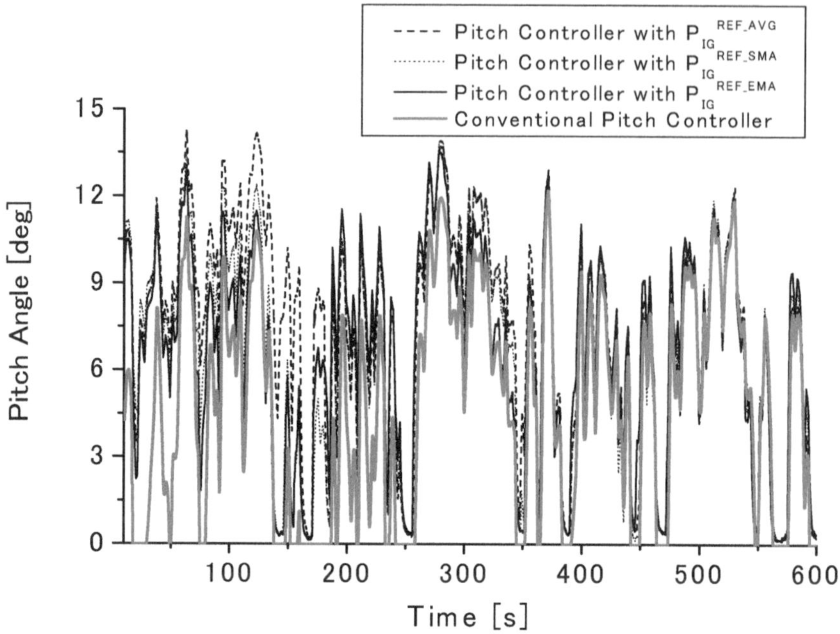

Fig. 3.42 Blade pitch angle of the wind turbine (Case 2)

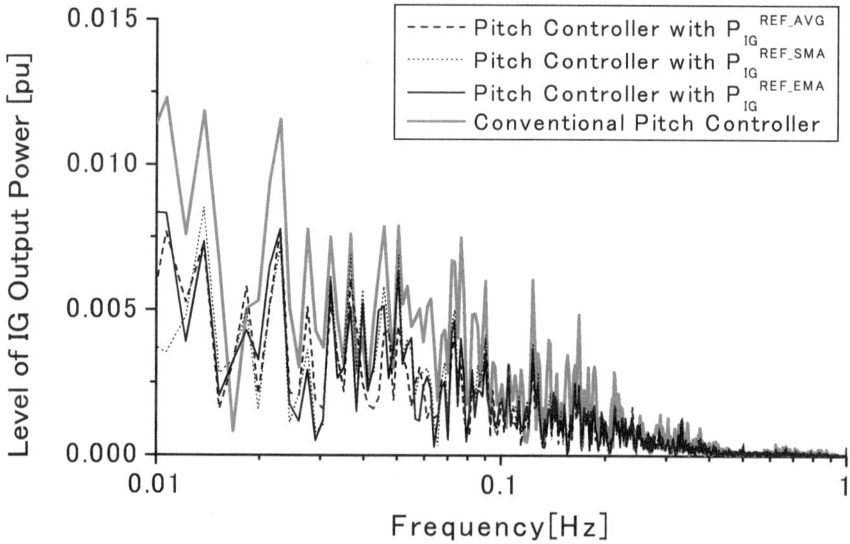

Fig. 3.43 Frequency spectrum of the IG output (Case 2)

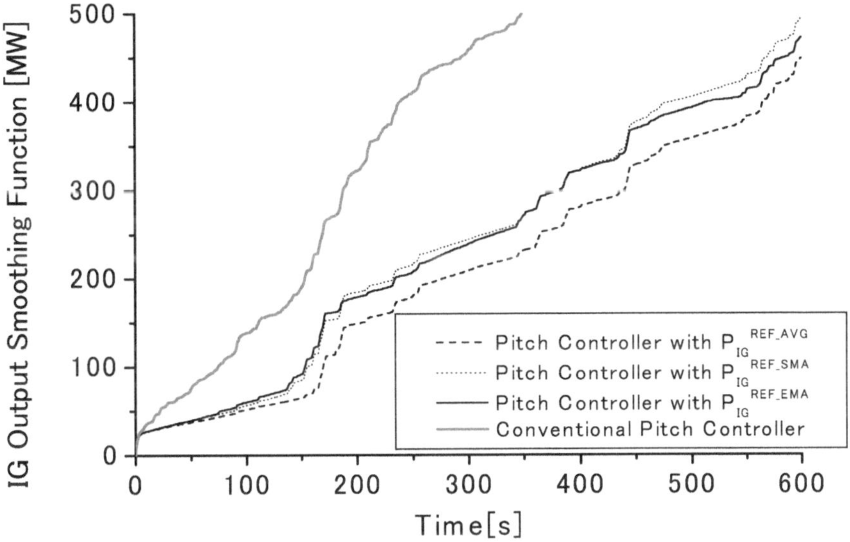

Fig. 3.44 Power-smoothing index of the induction generator (Case 2)

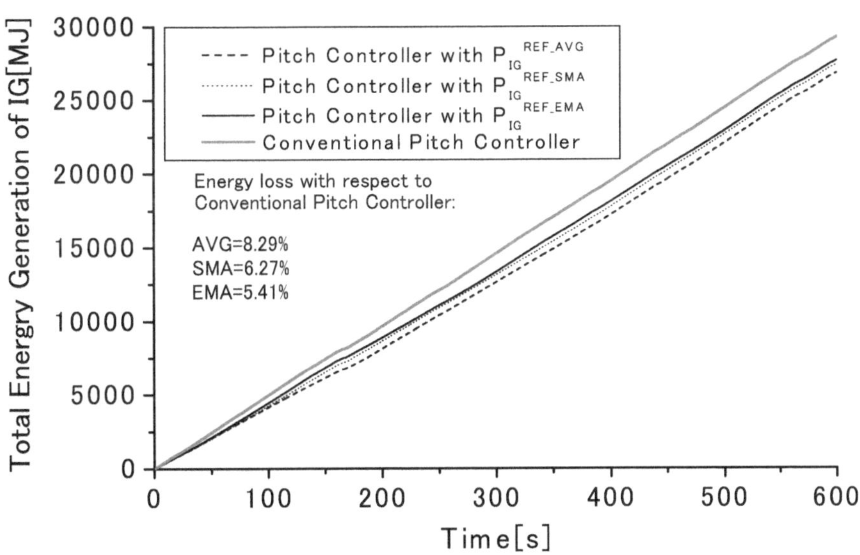

Fig. 3.45 Total energy generation by the induction generator (Case 2)

3.3.4.3 Case 3

In this case, the low wind speed pattern shown in Fig. 3.46 is used. The responses of the IG real power, the controller input power command, the blade pitch angle, the frequency spectrum of the IG output, the power-smoothing function, and the IG total energy generation are presented in Figs. 3.47 – 3.52, respectively.

It is clear that a FLC controlled pitch controller can smooth the IG output well even when the wind speed is low. But in this case, some points are noticeable. When the wind speed starts to increase rapidly at time 160 sec from a low value to a high value, $P_{IG}^{REF\_AVG}$ becomes zero around 245 sec, as shown in Fig. 3.48. Because at low wind speed, the average value of the turbine captured power is low and a big deviation can make the $P_{IG}^{REF\_AVG}$ zero. This can be understood from Eq. 3.6 and Fig. 3.29. But $P_{IG}^{REF\_SMA}$ and $P_{IG}^{REF\_EMA}$ always update themselves at the next period. Therefore, such situations can be avoided at low wind speed by using $P_{IG}^{REF\_SMA}$ or $P_{IG}^{REF\_EMA}$, as the controller input power command.

Moreover, the pitch controller with $P_{IG}^{REF\_AVG}$ gives more oscillation and more energy loss in the IG output power at low wind speed compared to those of $P_{IG}^{REF\_SMA}$ or $P_{IG}^{REF\_EMA}$, as shown in Figs. 3.50 and 3.52, respectively. Again the pitch controller command of $P_{IG}^{REF\_EMA}$ gives less oscillation in the IG output at low wind speed compared to that of $P_{IG}^{REF\_SMA}$. The overall smoothness is also better for $P_{IG}^{REF\_EMA}$ than that of $P_{IG}^{REF\_SMA}$, which is the key point of this analysis. Another point is that, when the wind speed suddenly increases or decreases around 170 to 300 sec, $P_{IG}^{REF\_EMA}$ can follow the trend more quickly than $P_{IG}^{REF\_SMA}$, as shown in Fig. 3.48. This is explained in Sect. 3.3.1.

The mechanical dead time shown in Table 3.6 is also large for $P_{IG}^{REF\_EMA}$ compared to $P_{IG}^{REF\_SMA}$, which reduces the mechanical load on the turbine blades.

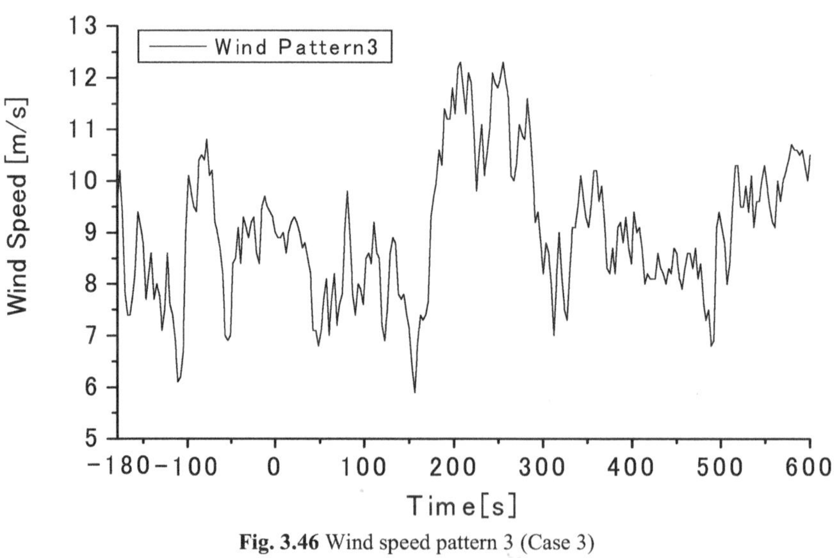

Fig. 3.46 Wind speed pattern 3 (Case 3)

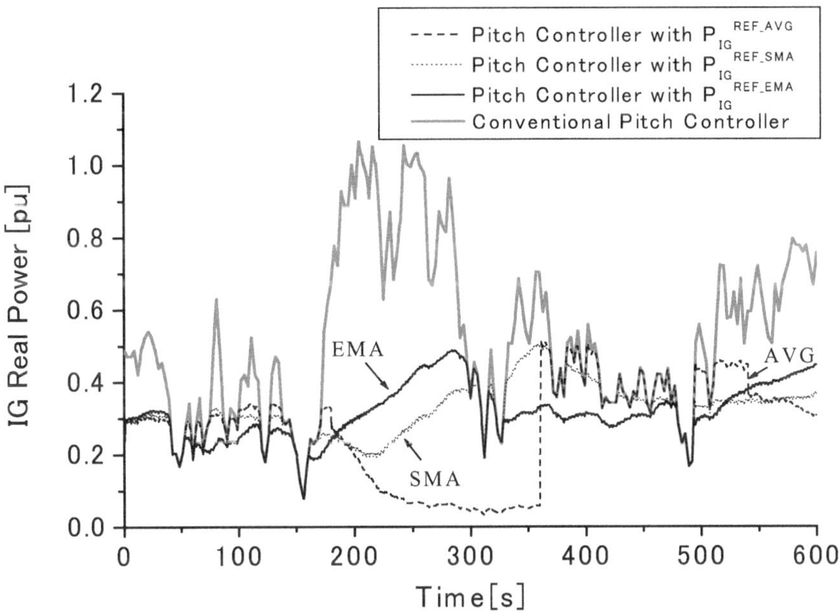

Fig. 3.47 Real power of the induction generator (Case 3)

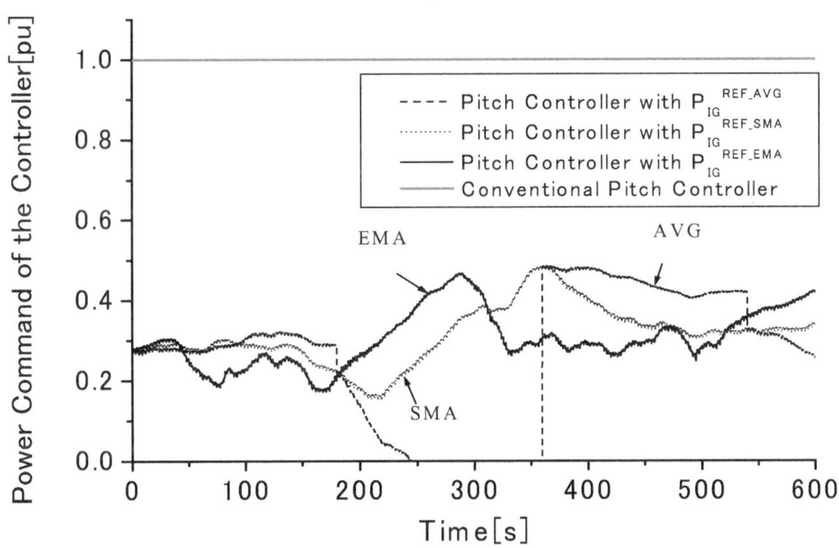

Fig. 3.48 Pitch controller power input command (Case 3)

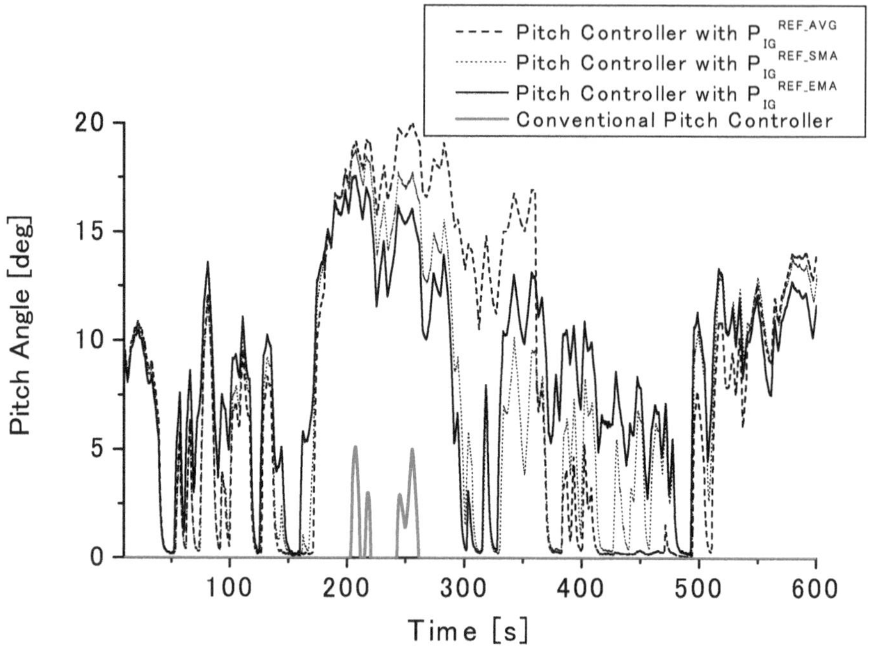

Fig. 3.49 Blade pitch angle of the wind turbine (Case 3)

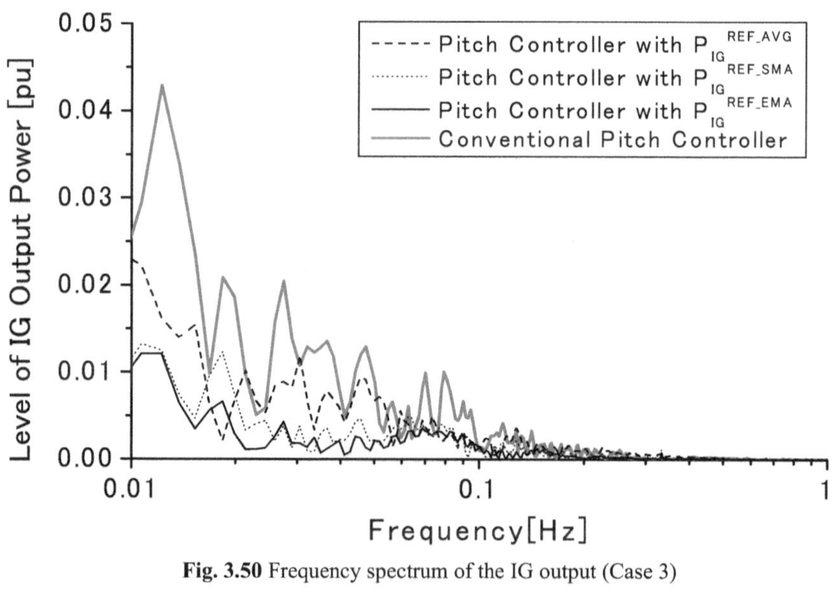

Fig. 3.50 Frequency spectrum of the IG output (Case 3)

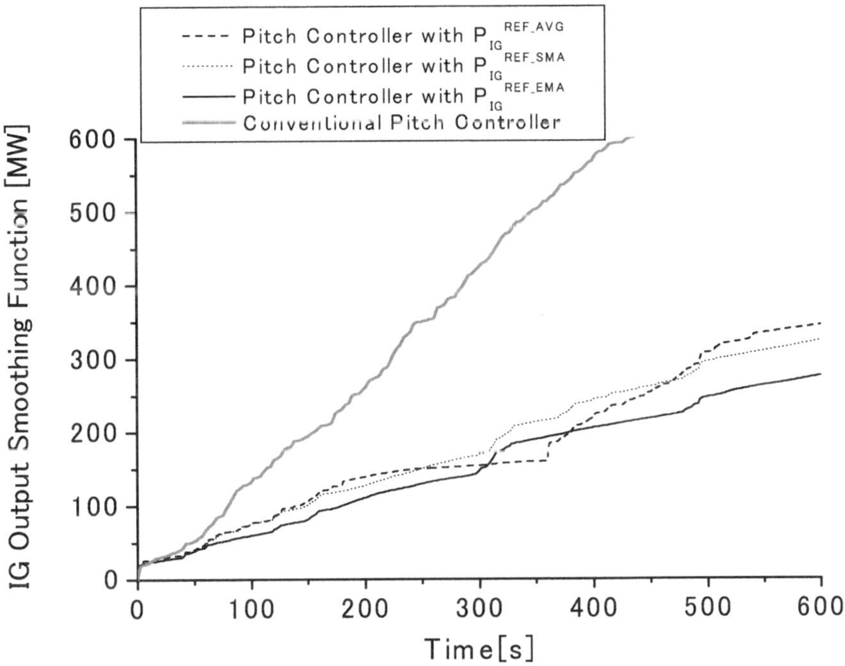

Fig. 3.51 Power-smoothing index of the induction generator (Case 3)

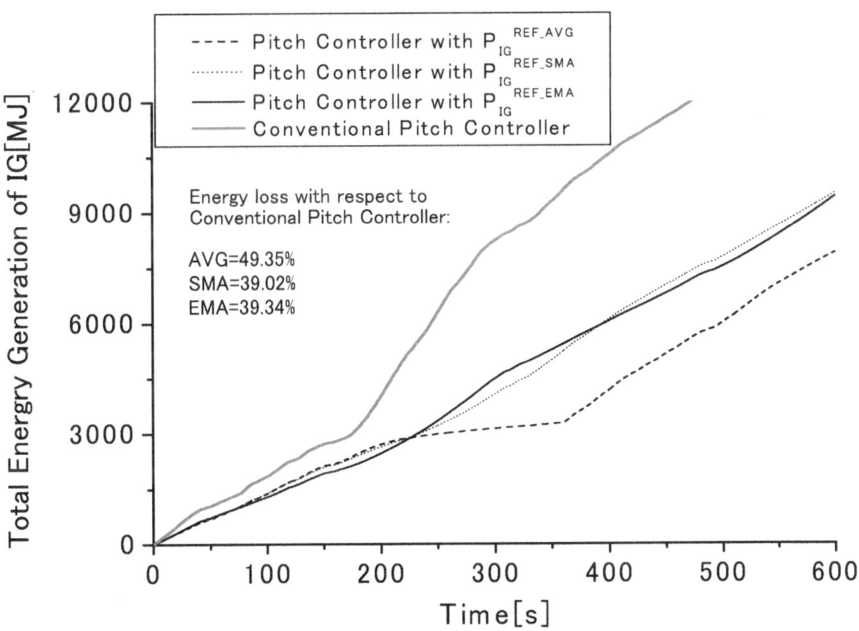

Fig. 3.52 Total energy generation by the induction generator (Case 3)

3.4 Chapter Summary

In this chapter, first, a logical pitch controller equipped with a FLC is presented in Sect. 3.2, which can maintain the output power of the wind generator at the rated level when the wind speed is over the rated speed. It can work well even when the wind speed is very high or fluctuates more frequently. Moreover, the same controller can enhance the transient stability during severe network disturbances in any wind condition. Using wind generator terminal voltage as the pitch controller input for robustness of the controller is emphasized also. The mechanical dead zone is considered in the simulations to obtain a realistic response. Simulation results show that the proposed pitch controller with the FLC unit gives better performance compared to that with a conventional PI unit. Therefore, using the FLC unit instead of the PI unit as the control strategy of the proposed logical pitch controller is recommended.

In Sect. 3.3, power smoothing of the wind generator by using a pitch controller is proposed. Nowadays, because most of the wind turbines are equipped with pitch controllers, this new feature of the pitch controller may receive much attention in the near future due to its cost-effectiveness. In Sect. 3.3, it is reported that the proposed pitch controller can smooth the wind power fluctuation well without using any energy storage systems. Therefore, the installation and maintenance costs can be significantly reduced. FLC is proposed as the control methodology of the pitch controller for wind power smoothing. Three different types of wind speed patterns are used to validate the effectiveness of the proposed pitch controller. Three different types of average values are adopted to generate the pitch controller input power command. It is reported that the controller input power command generated from the EMA can follow the wind speed trend well compared to those of SMA and AVG. Considering all operating conditions, it is recommended to use the EMA to generate a controller input power command from the viewpoint of lower energy loss and better smoothness. Some mechanical aspects regarding the controller design phase, which make the pitch controller practically applicable, are also considered throughout the simulations. Finally, it can be concluded that our proposed FLC based pitch controller can smooth the wind power fluctuation well.

Acknowledgements Special thanks to Mr. Hirotaka Kinoshita for his great effort to edit this entire chapter.

Chapter 4

STATCOM

Recently, various voltage-source or current-source inverter based FACTS devices have been used for flexible power flow control, secure loading, damping of power system oscillation, and even for stabilizing the wind generator. In this chapter, we propose the static synchronous compensator (STATCOM) based on a voltage source converter (VSC) PWM technique to stabilize a grid connected wind turbine generator system (WTGS). The three-level STATCOM topology is considered instead of the two-level STATCOM, as it is suitable for high-voltage application. The well-known cascade control scheme is used as the control strategy of the VSC based STATCOM. A multi-mass shaft model of a wind turbine generator system is also considered because shaft modeling has a big influence on the transient performance of a WTGS. Moreover, it is reported that a STATCOM can reduce the voltage fluctuation significantly in steady state operation under randomly varying wind speed conditions.

Another interesting feature of this chapter is the inclusion of the blade and shaft torsional oscillations of the wind turbine generator system during a network disturbance in the power system. Many reports are available in the power system literature where the damping of shaft torsional oscillations of the steam turbine generator system is discussed. Though a huge number of wind generators are going to be connected to the existing network, the damping of blade-shaft torsional oscillations of a WTGS has not yet been reported in the literature. In this chapter, a VSC based three-level STATCOM is proposed for damping of blade-shaft torsional oscillations of a WTGS. The six-mass drive train model of a WTGS is used for the sake of precise analysis. The damping performance of a STATCOM is compared with that of pitch controller because a pitch controller is attached to most recent wind turbines.

In all cases, both symmetrical and unsymmetrical faults are considered as network disturbances. Simulation results clearly show that a STATCOM can enhance the transient stability of a wind generator and can significantly minimize the blade-shaft torsional oscillation of a WTGS.

4.1 STATCOM Basics

The reactive power of a static synchronous compensator (STATCOM) is produced by means of power electronic equipment of the voltage source converter (VSC) type. The VSC converts the DC voltage into a three-phase set of output voltages with desired amplitude, frequency, and phase. The VSC may be of the two-level or the three-level type depending on the required output power and voltage. The schematic diagram of a STATCOM is shown in Fig. 4.1.

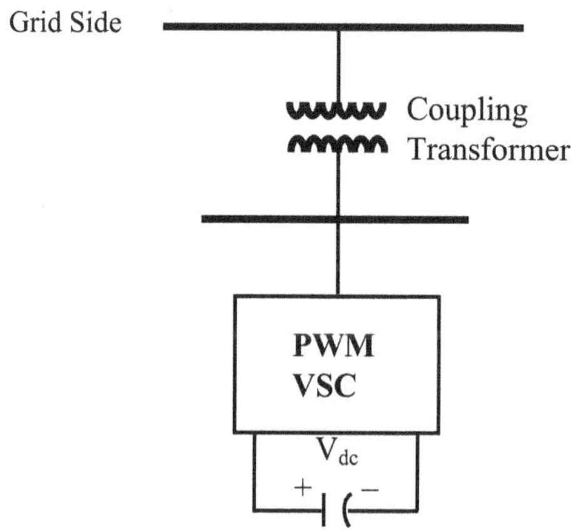

Fig. 4.1 Schematic diagram of the STATCOM

The name STATCOM is an indication that it has a characteristic similar to the synchronous condenser, but as an electronic device it has no inertia and is superior to the synchronous condenser in several ways, such as better dynamics, lower investment cost, and lower operating and maintenance costs [131].

With the advent of the STATCOM, still better performance can be reached in areas such as

- Dynamic voltage control in transmission and distribution systems
- Power oscillation damping in power transmission systems
- Transient stability improvement of power systems
- Ability to control not only reactive power but, if needed, also active power (with a DC energy source available).

STATCOM also brings further benefits such as

- A small footprint, due to the replacing of passive banks by compact electronic converters
- Modular, factory built equipment, reducing site works and commissioning time
- Use of encapsulated electronic converters, which minimizes environmental impact on the equipment.

Figures 4.2 and 4.3 show the "PCS 6000 STATCOM" front closed and open view, respectively. The "PCS 6000 STATCOM" is an ABB product rated at 12.5 MVAR[1].

Fig. 4.2 PCS 6000 STATCOM front closed view (Source: ABB (copyrights for the picture remain at ABB))

Fig. 4.3 PCS 6000 STATCOM front open view (Source: ABB (copyrights for the pictures remain at ABB))

[1] For detailed information on ABB products, visit at http://www.abb.com/

4.2 Model System

Figure 4.4 shows the model system used for the simulation analyses of the transient stability and blade-shaft torsional oscillation of the WTGS. Here, one synchronous generator (SG) is connected to an infinite bus through a transformer and a double circuit transmission line. One wind farm (induction generator, IG) is connected to the network via a transformer and short transmission line. In this analysis, an aggregated model of a wind farm is considered, where one large wind generator (50 MVA) represents several wind generators. A capacitor bank has been used for reactive power compensation at steady state, as described in Sect. 2.3.1 of Chap. 2. The STATCOM is connected to point K as shown in Fig. 4.4. The AVR (automatic voltage regulator) and GOV (governor) control system models for the synchronous generator described in Sect. 2.3.4.1 of Chap. 2 are used in this analysis. The generator parameters shown in Table 2.1 are used here. The system base is 100 MVA.

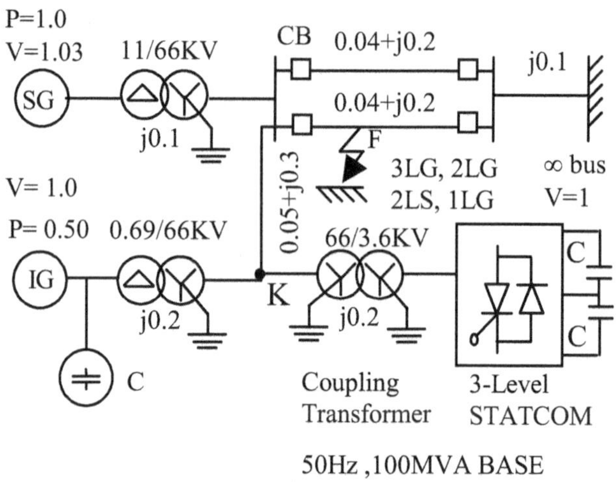

Fig. 4.4 Model system

The fixed speed WTGS is considered in this analysis. The wind turbine characteristics used in this case are described in Sect. 2.3.2 of Chap. 2. For the transient stability analysis, the two-mass shaft model is considered, which is found sufficient for the fault analysis of a WTGS, as reported in Sect. 2.3.3 of Chap. 2. The turbine inertia constant, H_{wt} (sec), generator inertia constant, H_g (sec), and the shaft stiffness, K_{2M} (pu) of the two-mass drive train of the WTGS are chosen as 3.0, 0.3 sec, and 90, respectively. On the other hand, for the blade-shaft torsional analysis the six-mass precise model is considered in this analysis. The six-mass drive train model parameters are shown in Table 2.4. All types of damping are neglected in this analysis to obtain the worst-case scenario.

4.3 Modeling and Control Strategy of STATCOM

Though the STATCOM configuration often consists of a two-level VSC, a DC energy storage device, and a coupling transformer connected in shunt with the ac system, the three-level STATCOM is considered in this book. To generate high output voltage, the VSC needs to use power semiconductor devices with high breakdown voltage. Due to the limitation of state-of-the-art semiconductor switch technology, the voltage rating is generally around 6 kV, with a mainstream switch voltage rating at 4.5 kV.

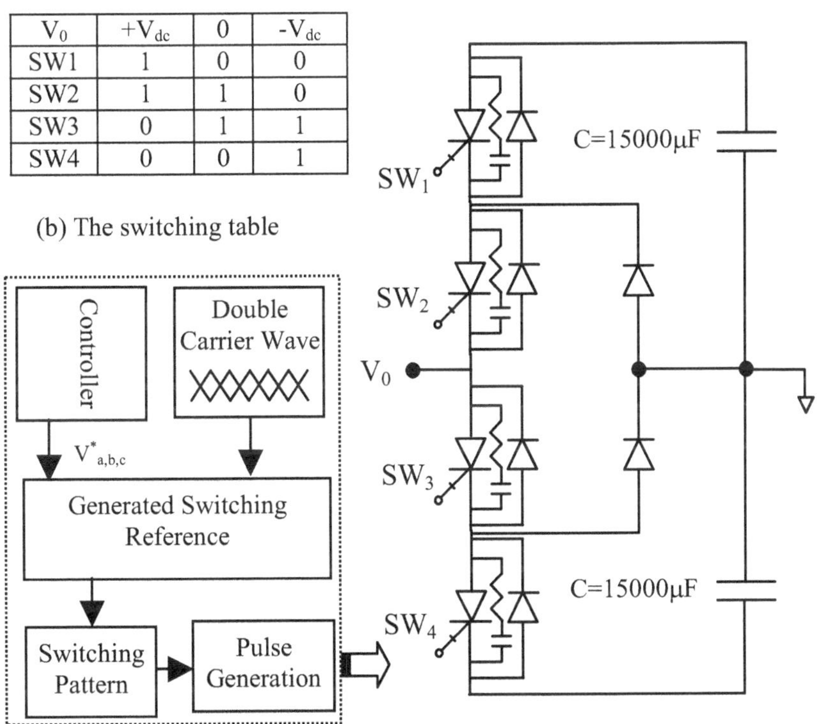

V_0	$+V_{dc}$	0	$-V_{dc}$
SW1	1	0	0
SW2	1	1	0
SW3	0	1	1
SW4	0	0	1

(b) The switching table

(c) Pulse generation system

(a) One pole structure

Fig. 4.5 Schematic diagram of a STATCOM switching circuit

There are two ways to increase the output voltage further; one is to use a series device connection, and the other is to use a multi-level converter. Although series connection of power semiconductor devices is a proven technology, there is still a restriction of the maximum allowable number of units. In this work, a three-level converter is used to increase the output voltage. The three-level converter has the advantages that the blocking voltage of each switching device is one half of the DC-link voltage in contrast to the full DC-link voltage in the case of the two-level converter. The harmonic contents of the three-level converter output voltage are much less than those of the two-level one, at the same switching frequency. The one pole structure of the three-level converter is shown in Fig. 4.5a. The GTO (gate turn-off thyristor) switching table and control methodology of the STATCOM are shown in Figs. 4.5b and c, respectively.

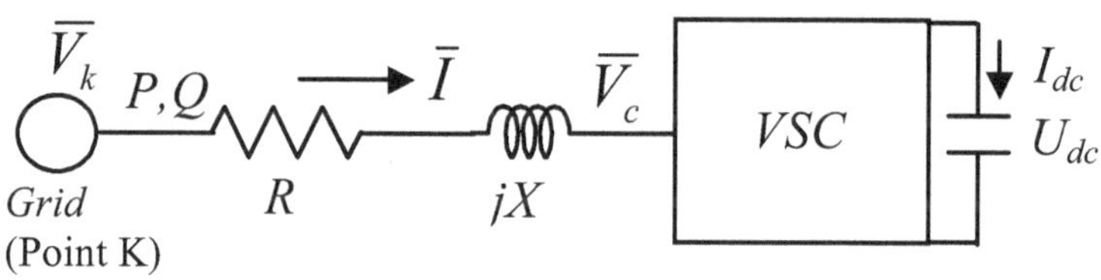

Fig. 4.6 Equivalent circuit of a VSC including a coupling transformer

The equivalent circuit of the VSC based STATCOM connected to a power grid through a coupling transformer can be expressed as shown in Fig. 4.6. The coupling transformer resistance and reactance are expressed by R and X, respectively. The phasor quantities $\overline{V}_k$, $\overline{V}_c$, and $\overline{I}$ represent the ac system voltage at point K of Fig. 4.4, output ac voltage of the converter, and the current following from the ac system to the converter, respectively. The real power, P, and the reactive power, Q, flowing through the converter can be derived easily as expressed below.

$$\left.\begin{aligned} P &\propto I_d \propto -V_{cq} \\ Q &\propto -I_q \propto -V_{cd} \end{aligned}\right\} \tag{4.1}$$

where, I_d and I_q are the d and q components of the current phasor, $\overline{I}$, and V_{cd} and V_{cq} are the d and q components of the VSC output ac voltage, $\overline{V}_c$.

The dq quantities and three-phase electrical quantities are related to each other by a reference frame transformation. The angle of the transformation is detected from the three-phase voltages (v_a,v_b,v_c) at point K of Fig. 4.4 by using a phase locked loop (PLL). The derivation of Eq. 4.1 is shown in the Appendix.

Again, if we neglect the switching losses and harmonics, then the following expression can be obtained:

$$P \propto U_{dc} I_{dc} \tag{4.2}$$

where, U_{dc} and I_{dc} are the DC voltage and current of the VSC, respectively.

From the above expressions, the well-known cascaded control scheme can be developed, as presented in Fig. 4.7. The VSC converts the DC voltage across the storage device into a set of three-phase ac output voltages. These voltages are coupled with the ac system through the impedance of the coupling transformer. Suitable adjustment of the phase and magnitude of the STATCOM output voltage allows effective control of power exchange between the STATCOM and the ac system.

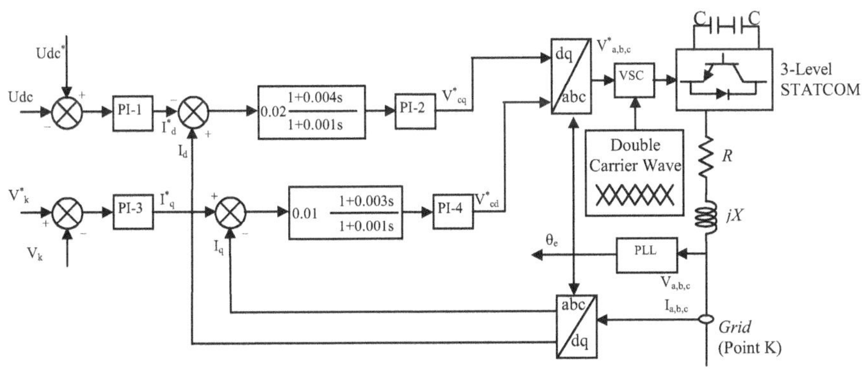

Fig. 4.7 Control block diagram of a three-level PWM based STATCOM

In this work, a STATCOM is used to regulate the wind generator terminal voltage. Usually, the wind farm terminal voltage is not kept at the rated voltage, but is reset to a desired value once or a few times a day. Because the proposed STATCOM can provide the necessary reactive power to wind generators, the wind farm terminal voltage can be kept at a desired reference level. The aim of the control is to maintain the voltage magnitude constant at the wind farm terminal, point K of Fig. 4.4, to which the STATCOM is connected. Therefore, an error signal is obtained by comparing the reference voltage with the rms value of the wind farm terminal voltage, V_k. A PI controller progresses this error signal and generates the command signal for the q-component of the VSC current. Next, three-phase VSC currents are detected and transformed to *dq* quantities. Finally, the second PI controller generates the *d*-axis reference voltage signal for the converter as it is proportional to the *q*-axis current, as shown in Eq. 4.1. Similarly, another two PI controllers work to generate the q-axis reference voltage signal to maintain a constant DC-link voltage. Then the three-phase reference signal is generated from the dq components of the reference voltage signals. These three-phase reference signals are compared with the double carrier wave signal in order to generate the switching signals for the GTO switched three-level VSC according to the switching rules mentioned in Fig. 4.5b.

High switching frequencies can be used to improve the efficiency of the converter, without incurring significant switching losses. In the simulation, the switching frequency chosen is 1000 Hz. The STATCOM rating is considered 50 MVA, which is the same as that of the wind farm. The snubber circuit resistance and capacitance values of the GTO devices in Fig. 4.5a are 5000 W and 0.05 mF, respectively. The DC-link voltage chosen is 6.6 kV. The STATCOM is connected to the 66 kV line through a single step-down transformer (66 kV/3.6 kV) with 0.2 p.u leakage reactance (base value 100 MVA). The DC-link capacitor value of each switching device is 15000 mF. The main power semiconductor devices incorporated in the converter design are gate turn-off thyristors (GTOs) rated at 6 kV. These devices are arranged to form a three-level VSC based converter circuit. The voltage stress of each switching device is clamped to one half of the DC-link voltage (3.3 kV), whereas the full DC-link voltage for a two level conventional converter is 6.6 kV. Thus the power devices could be fully utilized in the high-voltage range.

4.4 Simulation Analysis of a STATCOM Connected to a Fixed Speed Wind Generator

A symmetrical three-line-to-ground fault, 3LG, unsymmetrical double-line-to-ground fault, 2LG (phases B, C, and ground), line-to-line fault, 2LS (phase B and C), and single-line-to-ground fault, 1LG (phase C, and ground) are considered as network disturbance, which occurs at fault point F in Fig. 4.4. The fault occurs at 0.1 sec, the circuit breakers (CB) on the faulted lines are opened at 0.2 sec, and at 1.0 sec, are re-closed. The time step and simulation time have been chosen as 0.00002 sec and 10 sec, respectively. The simulation has been done by using PSCAD/EMTDC[2] [126].

---

[2] For the latest information on PSCAD/EMTDC, visit at http://pscad.com

4.4.1 Transient Stability Enhancement of WTGS by STATCOM

The initial values used in this analysis are shown in Table 4.1. The parameters of the PI controllers of the STATCOM shown in Fig. 4.7 are shown in Table 4.2.

Table 4.1 Initial conditions of generators and turbines used in Sect. 4.4.1

	SG	IG
P(pu)	1.0	0.50
V(pu)	1.03	0.999
Q(pu)	0.334	0.00 (0.239*)
E_{fd}(pu)	1.803	-
T_m(pu)	1.003	-
δ(deg)	50.71	-
slip	0.0	−1.09%
V_w (m/s)	-	11.797
β (deg)	-	0

* Reactive power drawn by induction generator.

Table 4.2 The parameters of the PI controllers used in Sect. 4.4.1

	PI-1	PI-2	PI-3	PI-4
K_p	2.0	1.8	2.0	0.03
T_i	0.10	0.002	0.20	0.002

4.4.1.1 Analysis Using One-Mass Lumped Model of WTGS

In this work, the responses of one large induction generator as an aggregated wind park model are presented. When a severe network disturbance occurs, the fixed capacitor bank with rated capacity value cannot compensate for the reactive power demand of the induction generator. Therefore, the induction generator terminal voltage falls, and the electromagnetic torque decreases suddenly. As a result, the induction generator goes out of step due to the large difference in mechanical and electromagnetic torques. But if a STATCOM is connected to the wind farm terminal, it can provide the necessary reactive power, and the induction generator doesn't go out of step.

Figures 4.8 and 4.9 show simulation results of induction generator terminal voltage and rotor speed, respectively, during a 3LG fault. They have been obtained using a one-mass lumped model of a WTGS. Figure 4.10 shows the reactive power output of the STATCOM. It is clear that a STATCOM can enhance the transient stability of a WTGS.

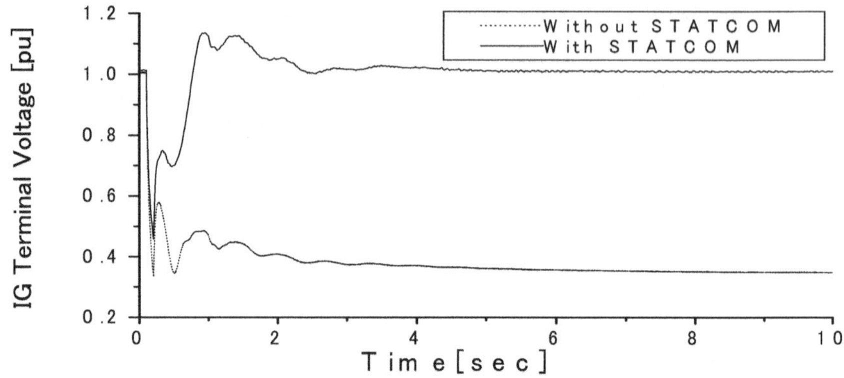

Fig. 4.8 Terminal voltage of induction generator with and without STATCOM (one-mass, 3LG)

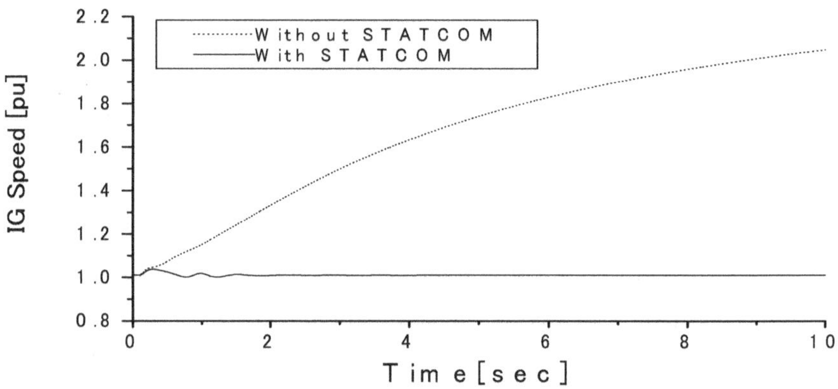

Fig. 4.9 Rotor speed of induction generator with and without STATCOM (one-mass, 3LG)

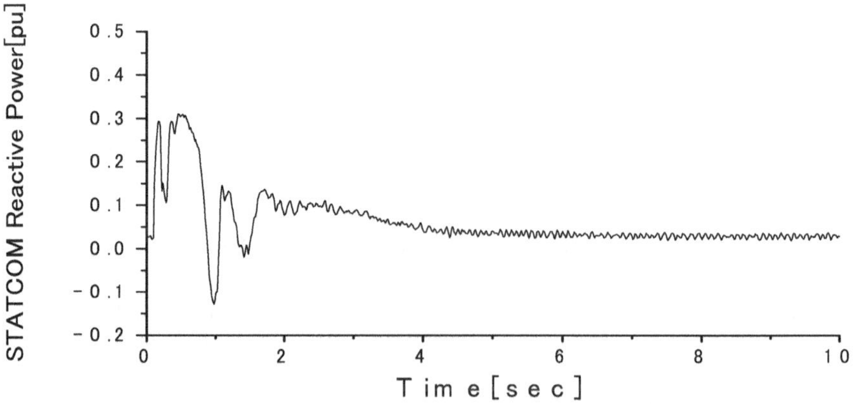

Fig. 4.10 Reactive power output of STATCOM (one-mass, 3LG)

4.4.1.2 Analysis Using Two-Mass Drive Train Model of WTGS

In Sect. 2.3.3 of Chap. 2, it is reported that drive train modeling has a significant effect on the transient stability of a fixed speed WTGS. In this section, it is investigated that the two-mass shaft model has a huge impact on the reactive power compensation of a WTGS by using a STATCOM. In Fig. 4.8, we observed that an IG can be compensated with reactive power by a STATCOM after a 3LG fault, even when the IG is generating rated output power (50 MW). But a one-mass lumped model is considered there. In this section, the two-mass shaft model is considered instead of the one-mass lumped model. The responses of the IG terminal voltage, IG rotor speed, turbine speed, and STATCOM reactive power output are shown in Figs. 4.11 – 4.14, respectively, for the symmetrical 3LG fault at fault point F of Fig. 4.4.

It is noticeable from Figs. 4.10 and 4.14 that the STATCOM needs to supply more reactive power for the two-mass drive train model compared to the one-mass lumped model to compensate for the reactive power demand of the IG when a network disturbance occurs in the power system.

The DC-link voltage of the VSC is shown in Fig. 4.15. The load angle of a synchronous generator is presented in Fig. 4.16. It is seen that a STATCOM can enhance the transient performance of a wind generator (for both models, one-mass and two-mass). In addition, it enhances the stability of the synchronous generator, i.e., the entire power system.

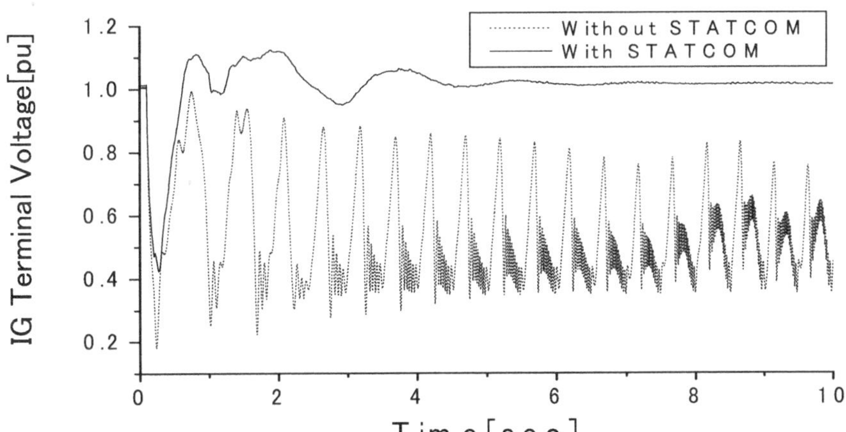

Fig. 4.11 Terminal voltage of induction generator with and without STATCOM (two-mass, 3LG)

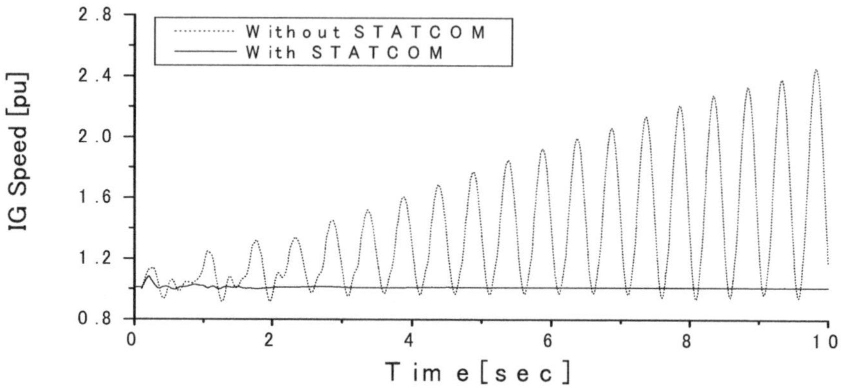

Fig. 4.12 Rotor speed of induction generator with and without STATCOM (two-mass, 3LG)

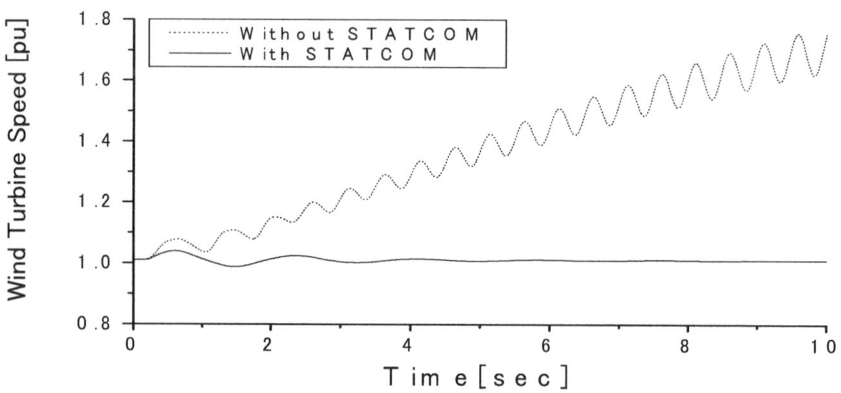

Fig. 4.13 Wind turbine speed with and without STATCOM (two-mass, 3LG)

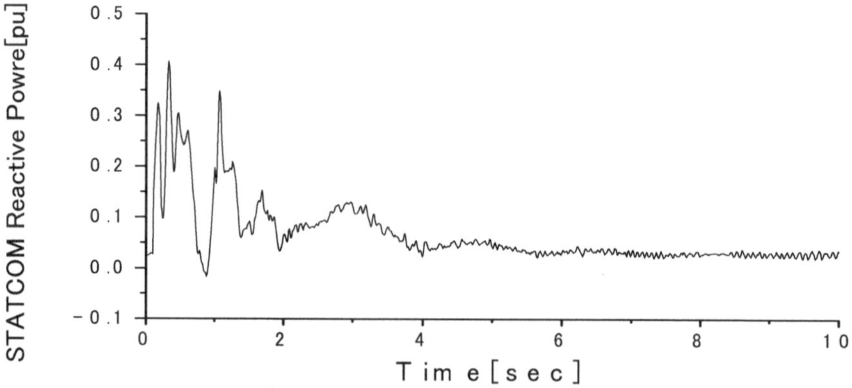

Fig. 4.14 Reactive power output of STATCOM (two-mass, 3LG)

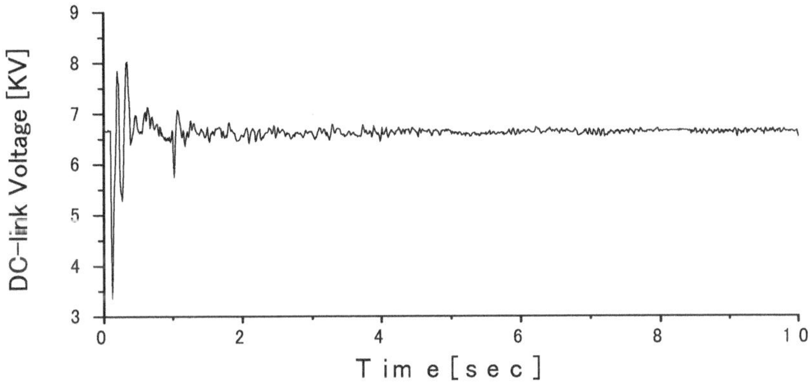

Fig. 4.15 DC-link voltage of STATCOM (two-mass, 3LG)

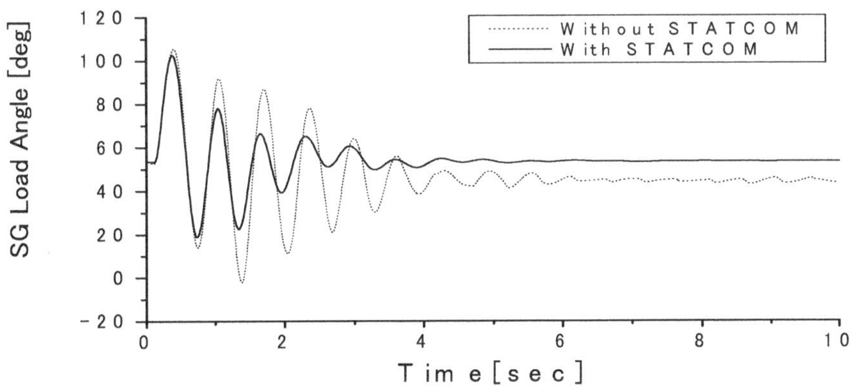

Fig. 4.16 Load angle of synchronous generator (two-mass, 3LG)

The responses of the IG terminal voltage, IG rotor speed, turbine speed, and STATCOM reactive power output are shown in Figs. 4.17 – 4.20, respectively, for the unsymmetrical 2LG fault. The load angle of a synchronous generator is presented in Fig. 4.21. For the 2LS fault, the responses of the IG terminal voltage, STATCOM reactive power, and load angle of SG are shown in Figs. 4.22 – 4.24, respectively. Finally, the responses of the IG terminal voltage, STATCOM reactive power, and load angle of the SG for a 1LG fault are shown in Figs. 4.25 – 4.27, respectively. It is seen from the simulation results that a STATCOM can enhance the transient stability of the induction and synchronous generators in both symmetrical and unsymmetrical fault conditions.

The parameters of the PI controllers of the VSC based three-level STATCOM are set based on the most severe three-line-to-ground fault. With those sets of parameters, the WTGS also becomes stable under other types of unsymmetrical fault conditions, which validates the sub-optimality of the constant setting.

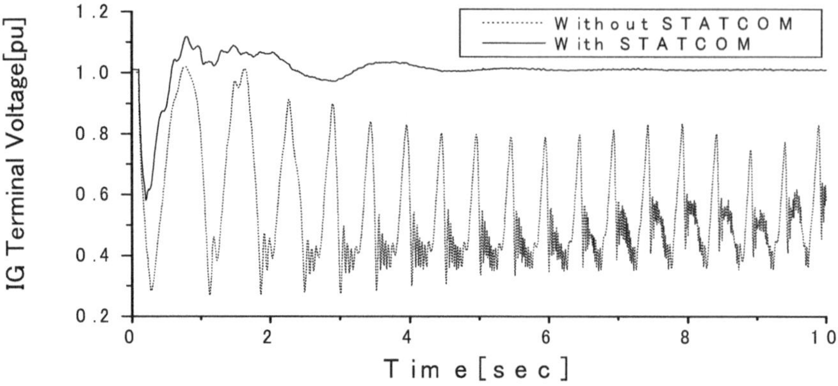

Fig. 4.17 Terminal voltage of induction generator with and without STATCOM (two-mass, 2LG)

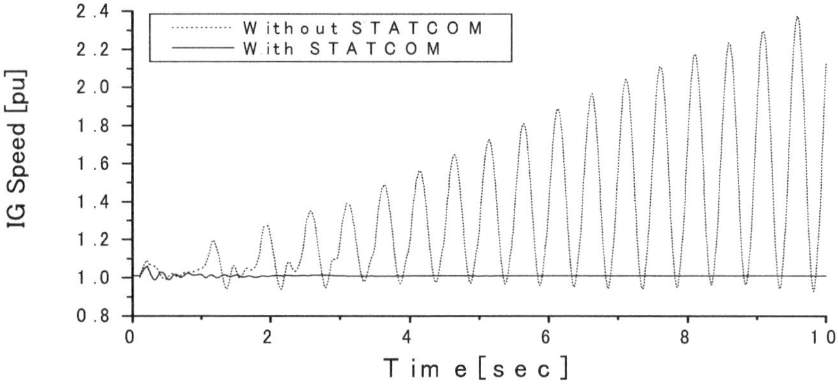

Fig. 4.18 Rotor speed of induction generator with and without STATCOM (two-mass, 2LG)

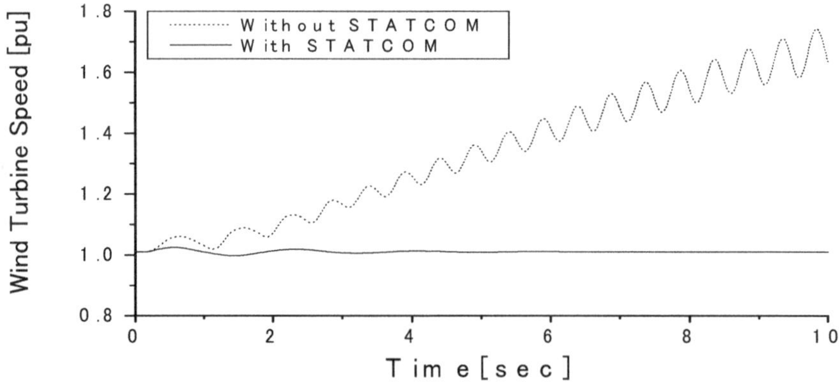

Fig. 4.19 Wind turbine speed with and without STATCOM (two-mass, 2LG)

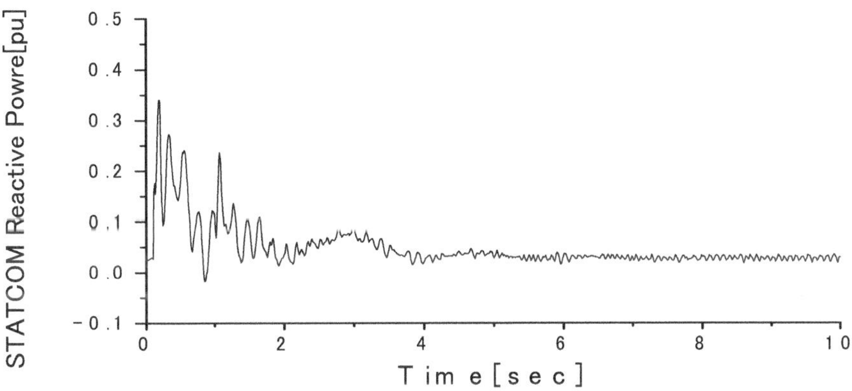

Fig. 4.20 Reactive power output of STATCOM (two-mass, 2LG)

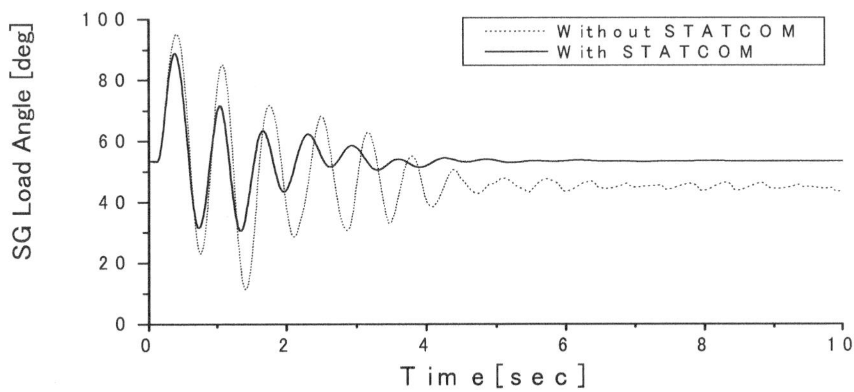

Fig. 4.21 Load angle of synchronous generator (two-mass, 2LG)

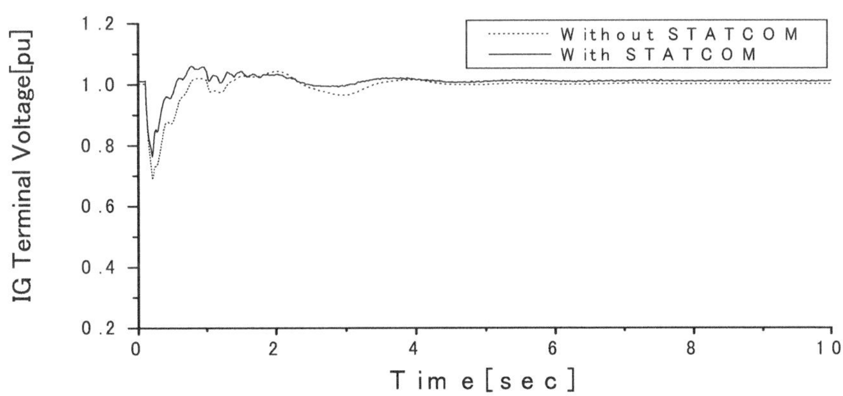

Fig. 4.22 Wind turbine speed with and without STATCOM (two-mass, 2LS)

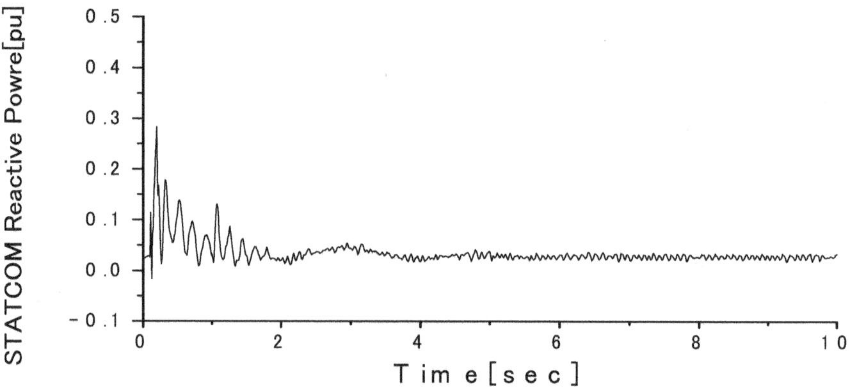

Fig. 4.23 Reactive power output of STATCOM (two-mass, 2LS)

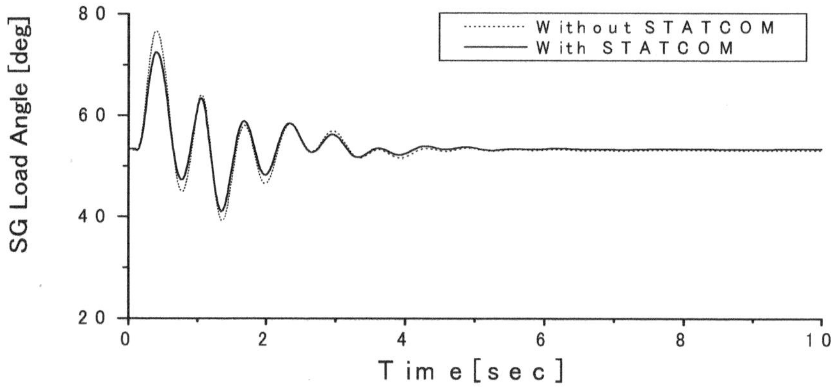

Fig. 4.24 Load angle of synchronous generator (two-mass, 2LS)

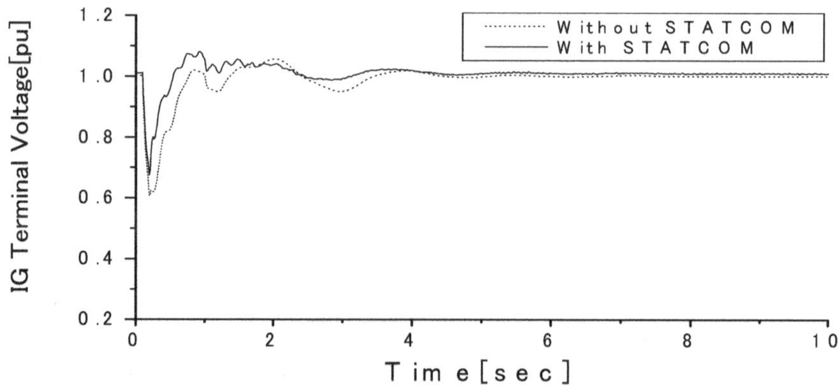

Fig. 4.25 Wind turbine speed with and without STATCOM (two-mass, 1LG)

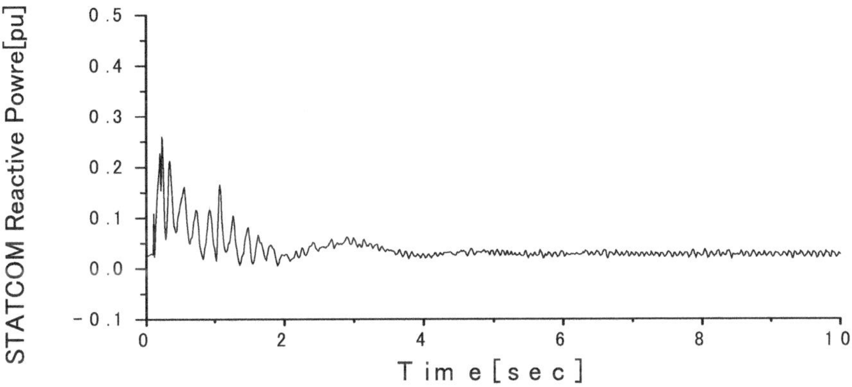

Fig. 4.26 Reactive power output of STATCOM (two-mass, 1LG)

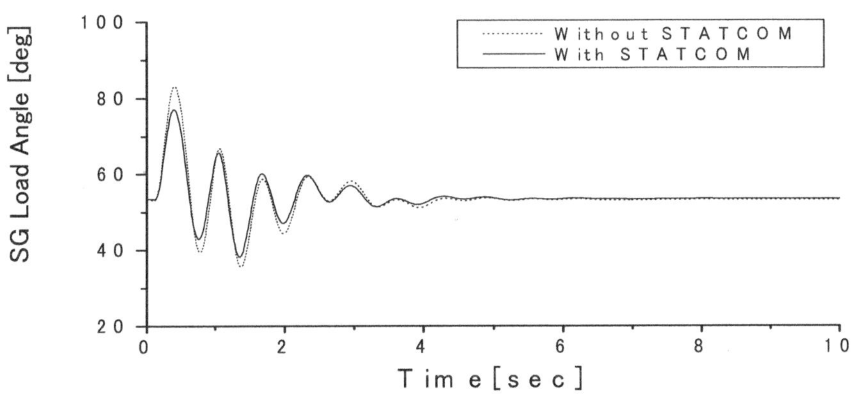

Fig. 4.27 Load angle of synchronous generator (two-mass, 1LG)

4.4.2 Power Quality Improvement of Wind Generator by STATCOM

Because wind speed is intermittent and stochastic by nature, the generated power from fixed speed wind generator fluctuates randomly. Therefore, the reactive power demand of the induction generator for unity power factor operation is not always constant. As a result, with only rated capacitor bank, the induction generator terminal voltage fluctuates randomly and the power quality of the wind generator deteriorates.

Using a STATCOM, this voltage fluctuation of a fixed speed wind generator under randomly varying wind speed can be reduced. For simulation analysis, the real wind speed data obtained on Hokkaido Island, JAPAN, shown in Fig. 4.28, are used. Figure 4.29 shows the terminal voltage response of the wind generator under that varying wind speed with and without the proposed VSC based three-level STATCOM. It is seen that a STATCOM can significantly reduce the wind generator voltage fluctuation, i.e., a STATCOM can improve the power quality of the wind generator.

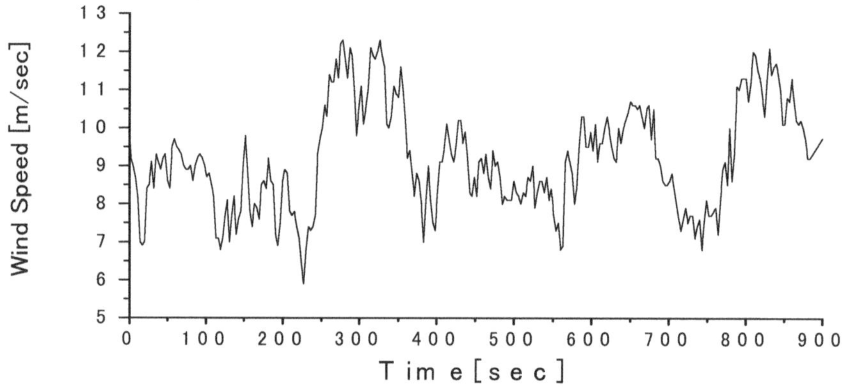

Fig. 4.28 Wind speed data measured on Hokkaido Island, Japan

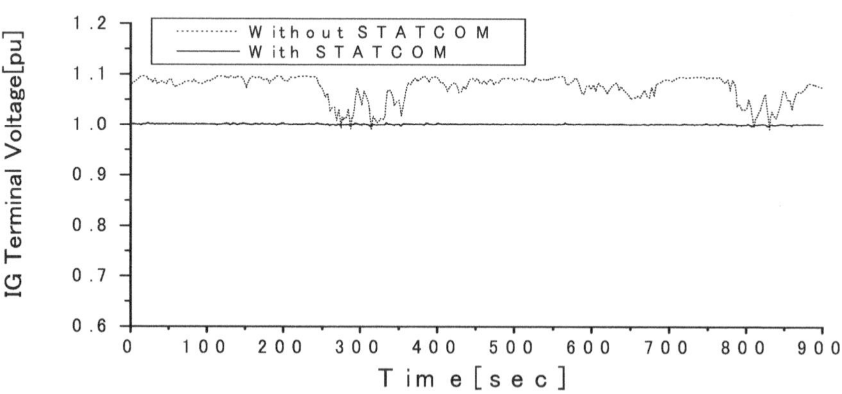

Fig. 4.29 Terminal voltage of induction generator with and without STATCOM

4.4.3 Damping of Blade-Shaft Torsional Oscillation of WTGS by STATCOM

Another objective of this work is to decrease the blade-shaft torsional oscillations of a fixed-speed WTGS under network disturbances. When a short-circuit fault occurs in the power system including a wind generator, the short-circuit current causes a voltage drop at the terminal of the wind generator. Therefore, the electromagnetic torque of the wind generator also drops suddenly because the electromagnetic torque is proportional to the terminal voltage of the wind generator. But the mechanical torque of the wind turbine doesn't change rapidly during that short time interval. As a result, the turbine and generator rotors will accelerate due to the large difference in mechanical and electromagnetic torques of the WTGS.

The electromagnetic torque of a wind generator can be restored quickly by recovering the terminal voltage as soon as possible. Due to the characteristic of the induction generator, a large reactive power is required to recover the air gap flux. Our control strategy is to sense the terminal voltage variations of the induction generator and then to provide the necessary reactive power from the three-level PWM based STATCOM. As a result, the terminal voltage can be returned to its pre-fault level, and the electromagnetic torque can also be restored quickly. Finally, damping of the blade-shaft torsional oscillations of a WTGS can also be achieved.

The six-mass drive train model can express the wind turbine generator system precisely. Therefore, in this analysis, we considered the six-mass drive train model. The initial values used in this analysis are shown in Table 4.3. The parameters of the PI controllers shown in Fig. 4.7 are given in Table 4.4.

Table 4.3 Initial conditions of generators and turbines used in Sect. 4.4.3

	SG	IG
P(pu)	1.0	0.45
V(pu)	1.03	1.02
Q(pu)	0.288	0.027 (0.222)*
Efd(pu)	1.76	-
Tm(pu)	1.003	-
δ (deg)	50.65	-
slip	0.0	−0.93%
Vw (m/s)	-	11.258
β (deg)	-	0

* Reactive power drawn by the induction generator

In the simulation study, all types of damping are neglected to investigate the worst-case scenario. The results with a STATCOM are compared with those pitch

Table 4.4. The parameters of the PI controllers used in Sect. 4.4.3

	PI-1	PI-2	PI-3	PI-4
K_p	3.0	0.10	3.0	0.1
T_i	0.25	0.004	0.25	0.002

controller and when no controller is used. The pitch controller used in this analysis is described in Sect. 3.2 of Chap. 3.

The responses of the real power, terminal voltage of induction generator, high-speed shaft torque, low-speed shaft torque, and the torque acting between hub and turbine blades, are shown in Fig. 4.30a, Fig. 4.31a, Fig. 4.32a, Fig. 4.33a, and Fig. 4.34a, respectively, for the unsymmetrical 2LG fault, where the pitch controller and the STATCOM are not considered. The responses for the 2LG fault with the pitch controller and the STATCOM are shown in Figs. 4.30b – 4.34b and Figs. 4.30c – 4.34c, respectively. For the 3LG fault, the responses without any controller, with the pitch controller, and with the STATCOM are shown in Figs. 4.35a – 4.39a, Figs. 4.35b – 4.39b, and Figs. 4.35c – 4.39c, respectively.

It is well known that symmetrical and unsymmetrical faults cause 50 Hz and 100 Hz torque oscillations in the induction generator that are transmitted to the drive train. Due to the relatively high spring constant of the high-speed shaft, it is expected that those oscillations will reach the gearbox. The symmetrical 3LG fault creates the maximum torque stress on the turbine shaft and blades as shown in Figs. 4.37a – 4.39a.

It is seen that the pitch controller can somewhat decrease the blade-shaft torsional oscillations of a WTGS for both 2LG and 3LG faults, as shown in Figs. 4.32b – 4.34b and Figs. 4.37b – 4.39b, respectively. The blade-shaft oscillations up to 4 sec in Figs. 4.32b – 4.34b and Figs. 4.37b – 4.39b may be due to the low shaft stiffness of the WTGS. The IG rotor and turbine speeds with the pitch controller are shown in Fig. 4.40 only for the 3LG fault. If a one-mass lumped model is used, such turbine and generator rotor speed oscillations cannot be obtained because low shaft stiffness is not considered in the lumped model. However, the STATCOM can significantly decrease the blade-shaft torsional oscillations of a WTGS for both 2LG and 3LG faults, as shown in Fig. 4.32c – 4.34c and Figs. 4.37c – 4.39c, respectively. Because the STATCOM can provide the necessary reactive power for the induction generator after a network disturbance, the terminal voltage of the wind generator can return to its pre-fault level, and the IG rotor and wind turbine can become stable.

The IG rotor and turbine speeds without and with the STATCOM are shown in Figs. 4.41 and 4.42, respectively for a 3LG fault. The reactive power supplied from the STATCOM is shown in Fig. 4.43 for a 3LG fault. From the simulation results, it is clear that a STATCOM can significantly decrease the blade-shaft torsional oscillations of a fixed speed WTGS during a network disturbance in the power system.

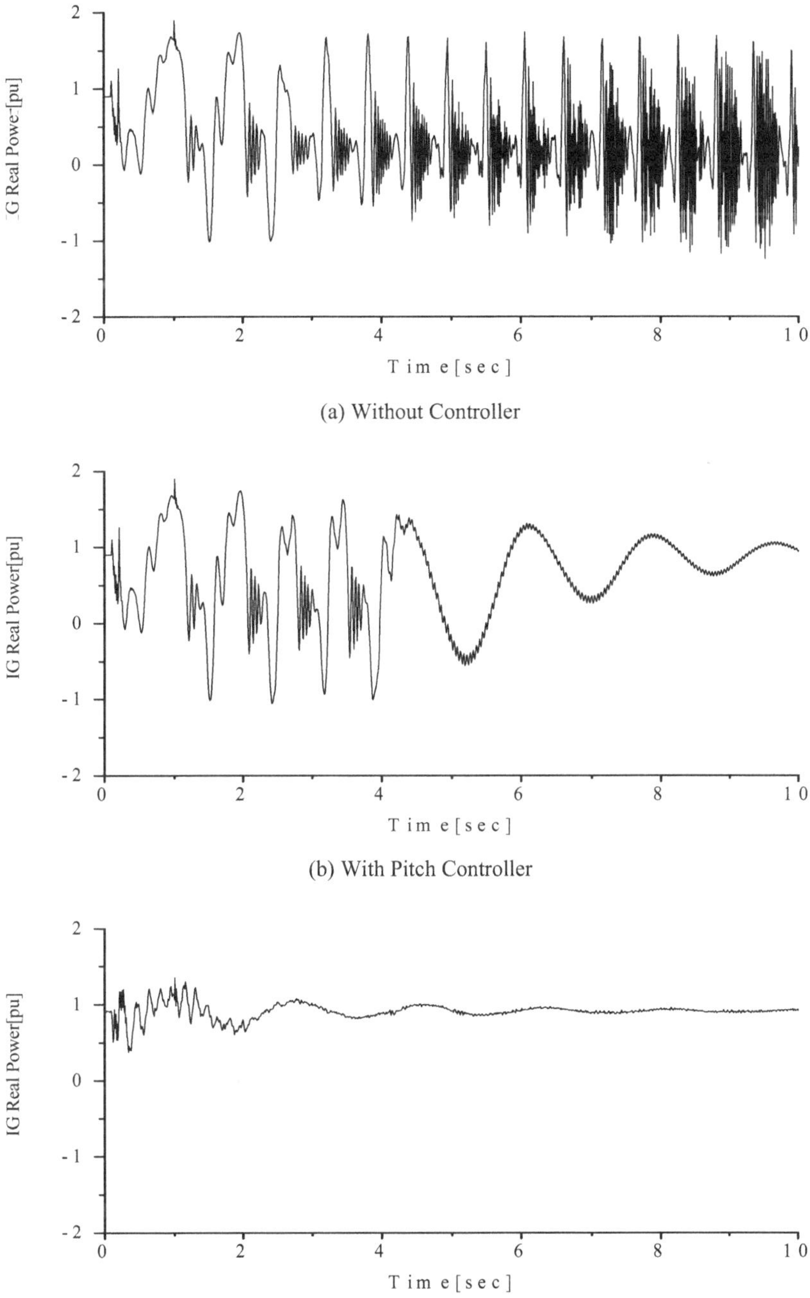

(a) Without Controller

(b) With Pitch Controller

(c) With STATCOM

Fig. 4.30 Real power of induction generator (unsymmetrical 2LG fault)

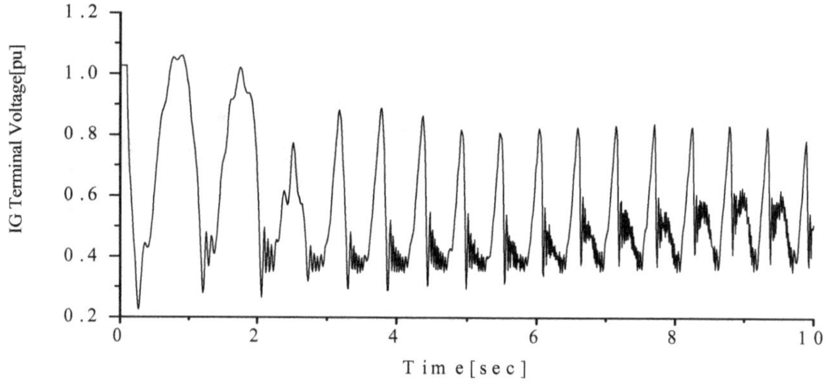

(a) Without Controller

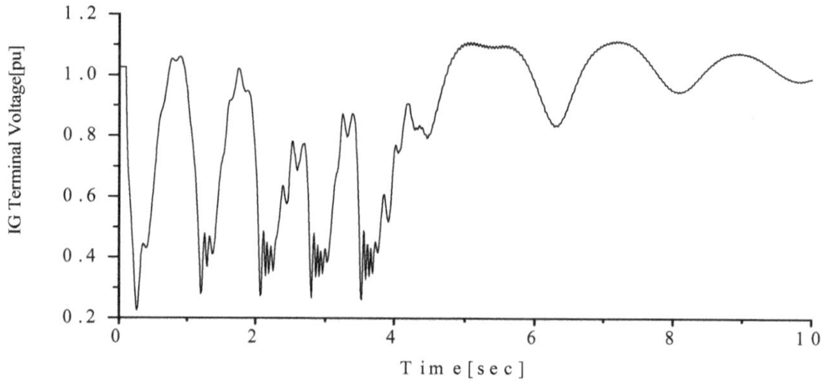

(b) With Pitch Controller

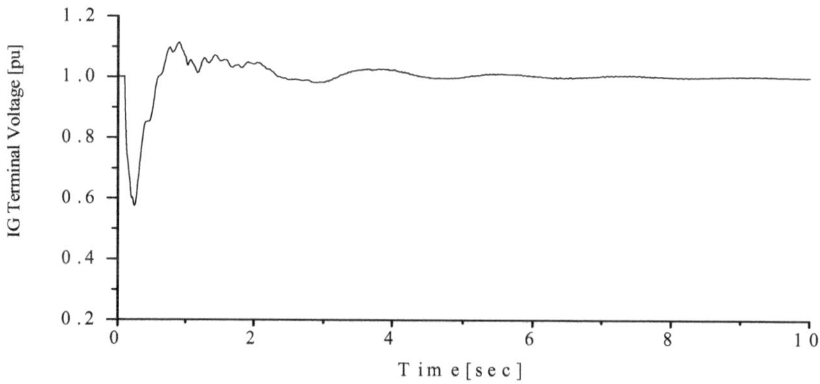

(c) With STATCOM

Fig. 4.31 Terminal voltage of induction generator (unsymmetrical 2LG fault)

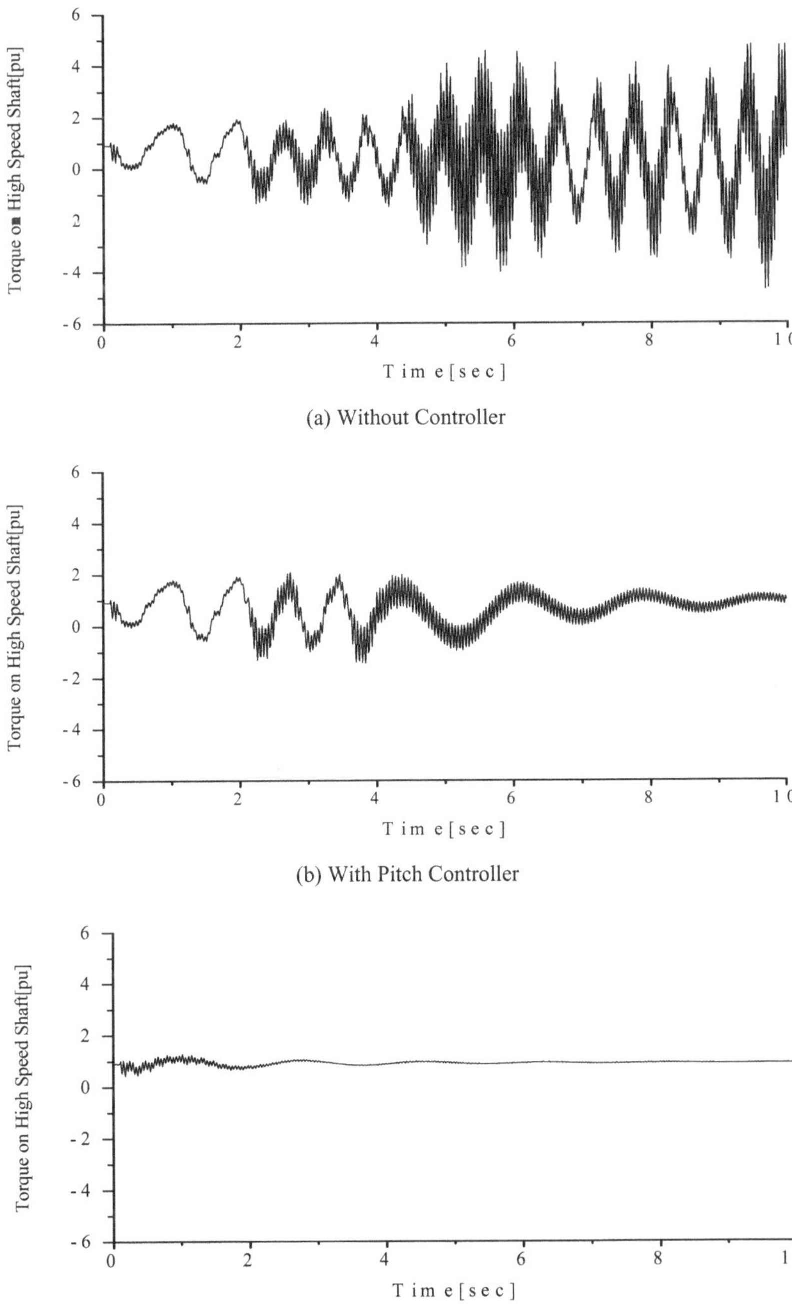

(a) Without Controller

(b) With Pitch Controller

(c) With STATCOM

Fig. 4.32 Torque on the high-speed side shaft of WTGS (unsymmetrical 2LG fault)

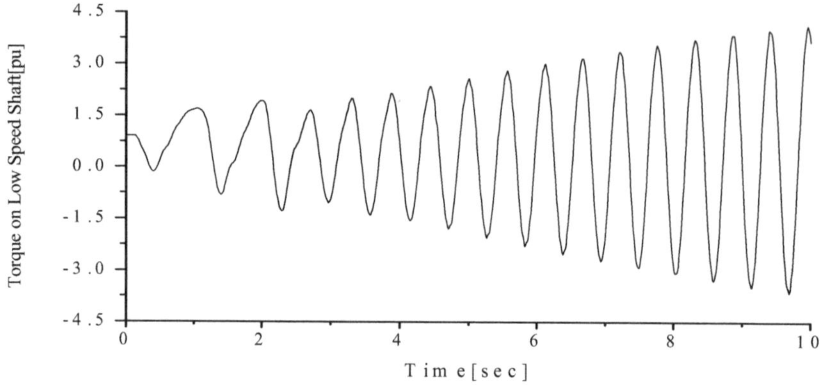

(a) Without Controller

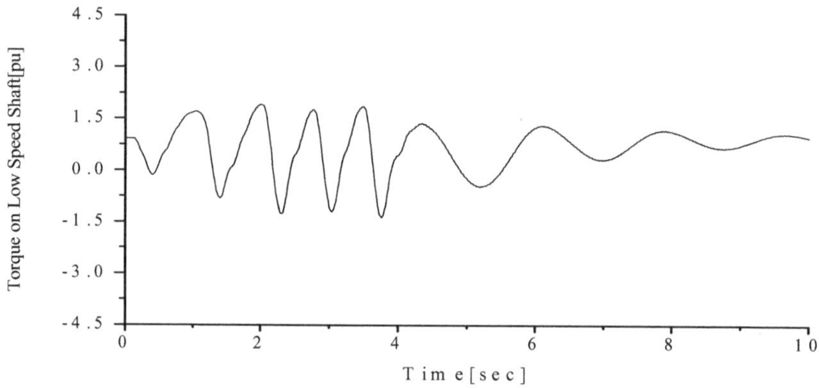

(b) With Pitch Controller

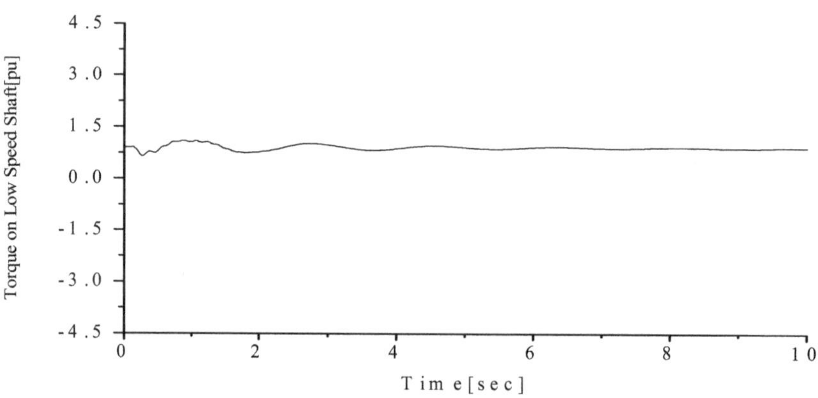

(c) With STATCOM

Fig. 4.33 Torque on the low-speed side shaft of WTGS (unsymmetrical 2LG fault)

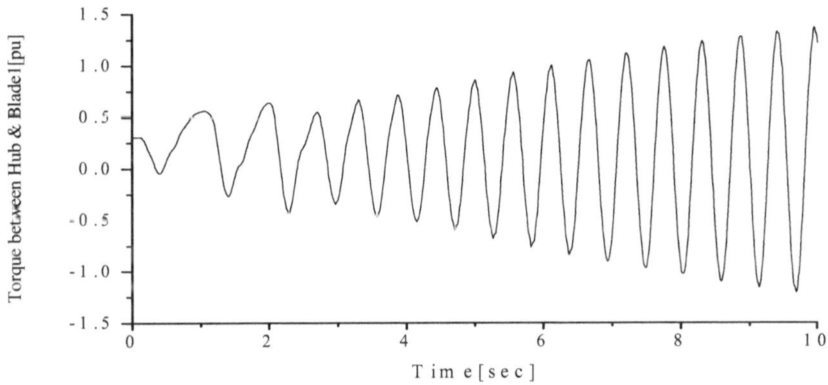

(a) Without Controller

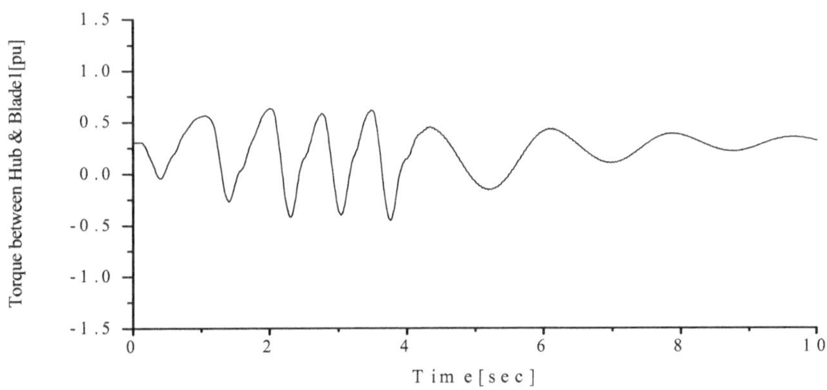

(b) With Pitch Controller

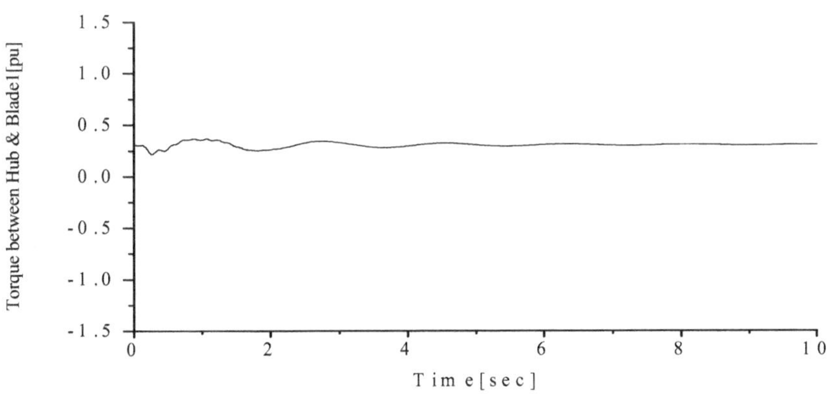

(c) With STATCOM

Fig. 4.34 Torque between hub and blade of WTGS (unsymmetrical 2LG fault)

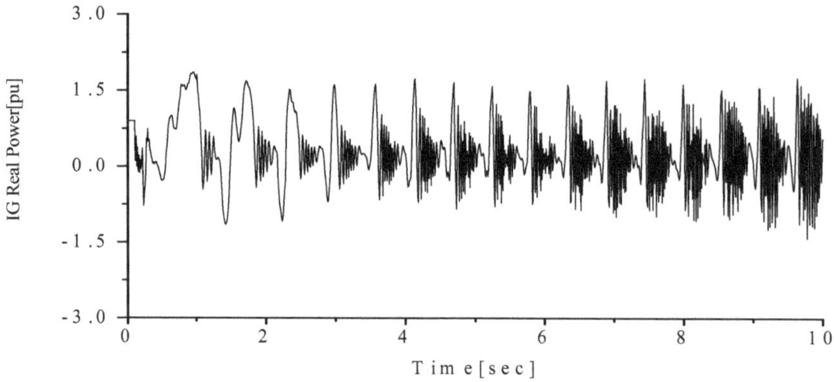

(a) Without Controller

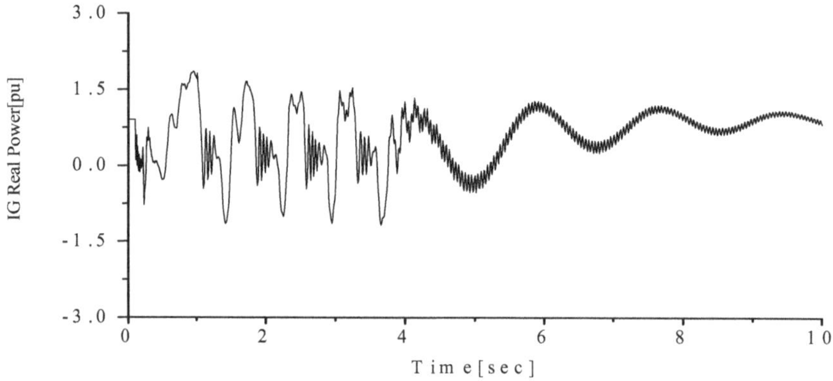

(b) With Pitch Controller

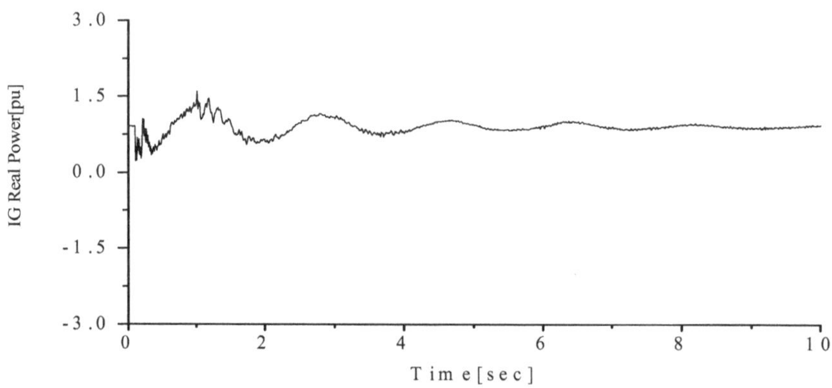

(c) With STATCOM

Fig. 4.35 Real power of induction generator (symmetrical 3LG fault)

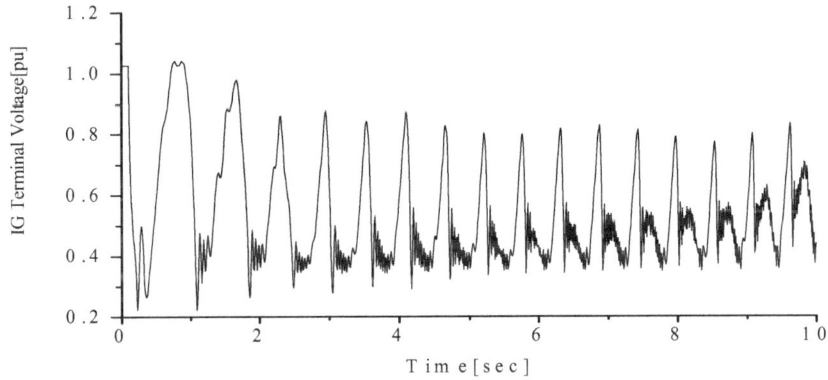

(a) Without Controller

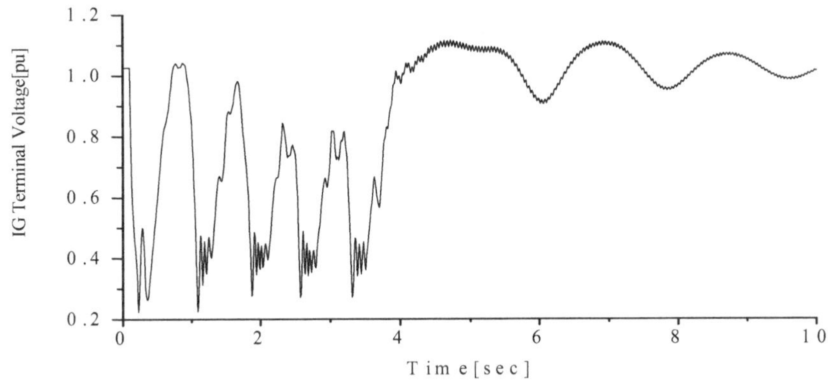

(b) With Pitch Controller

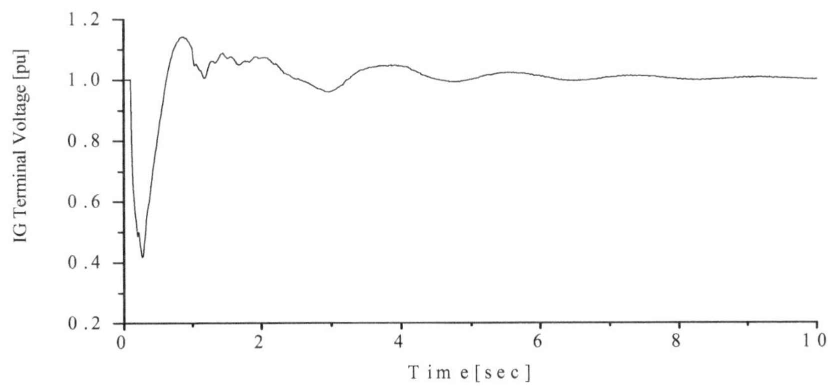

(c) With STATCOM

Fig. 4.36 Terminal voltage of induction generator (symmetrical 3LG fault)

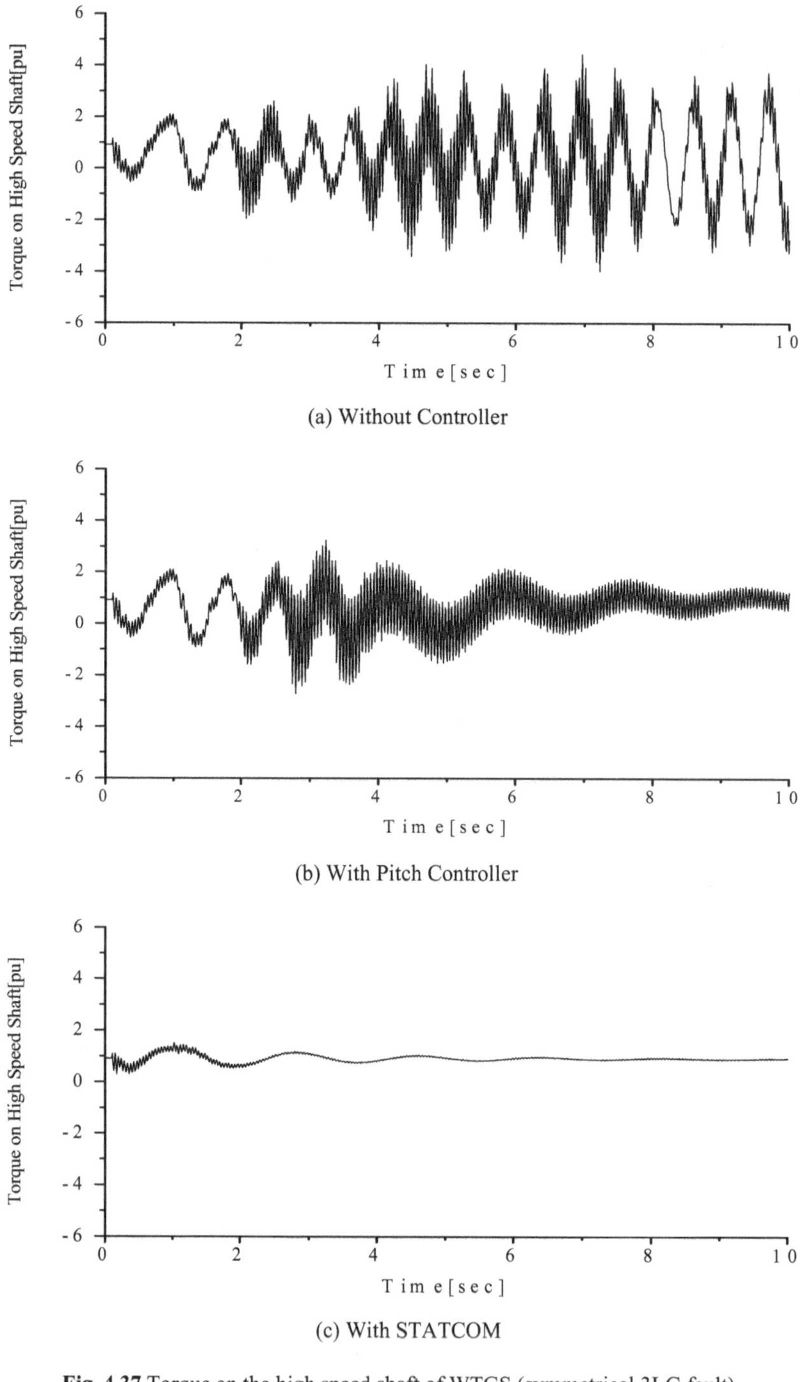

Fig. 4.37 Torque on the high speed shaft of WTGS (symmetrical 3LG fault)

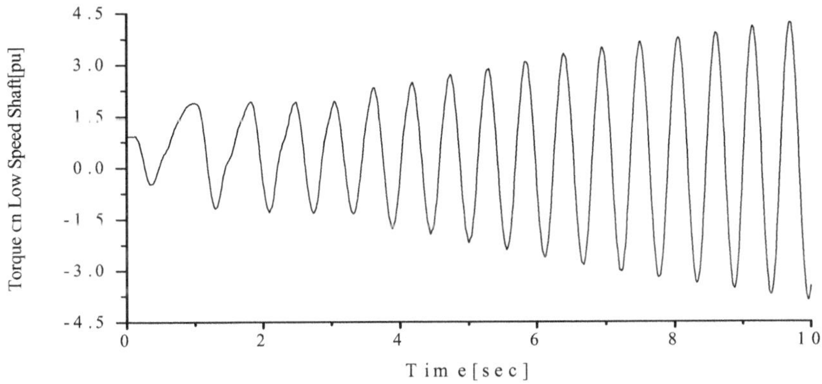

(a) Without Controller

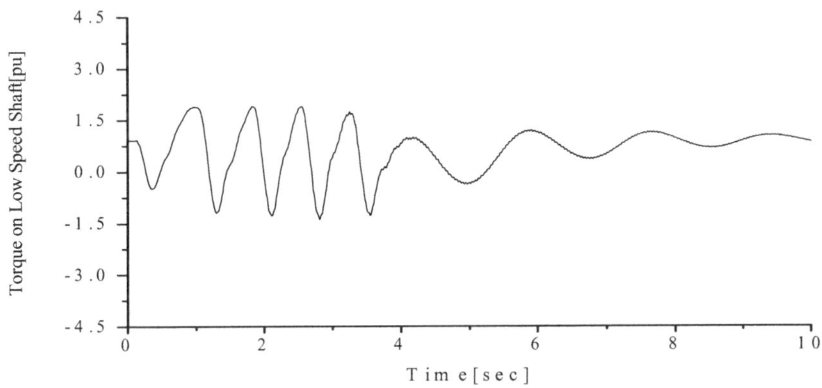

(b) With Pitch Controller

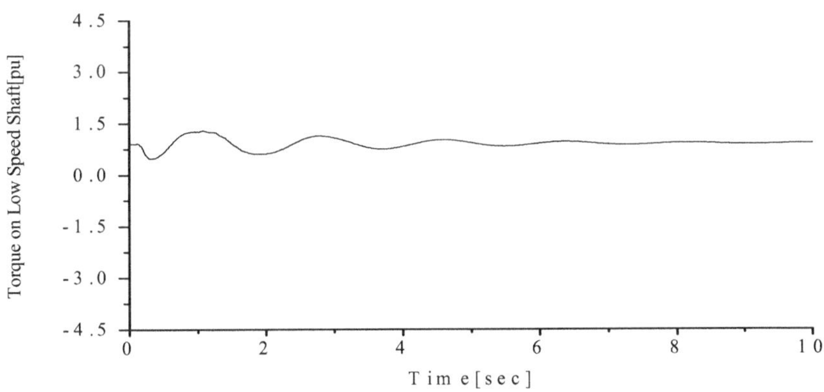

(c) With STATCOM

Fig. 4.38 Torque on the low speed shaft of WTGS (symmetrical 3LG fault)

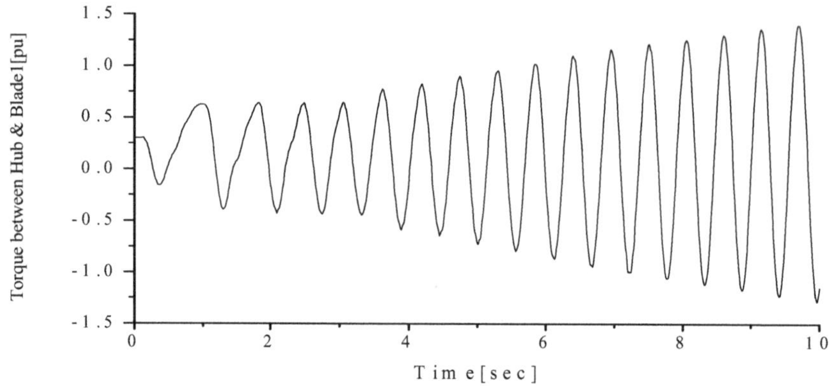

(a) Without Controller

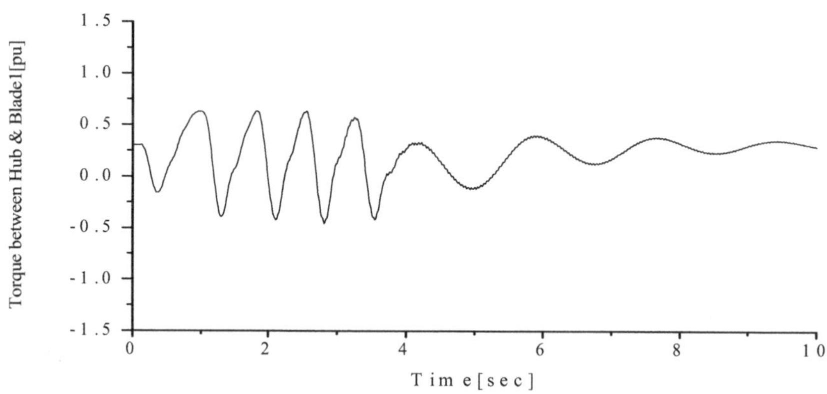

(b) With Pitch Controller

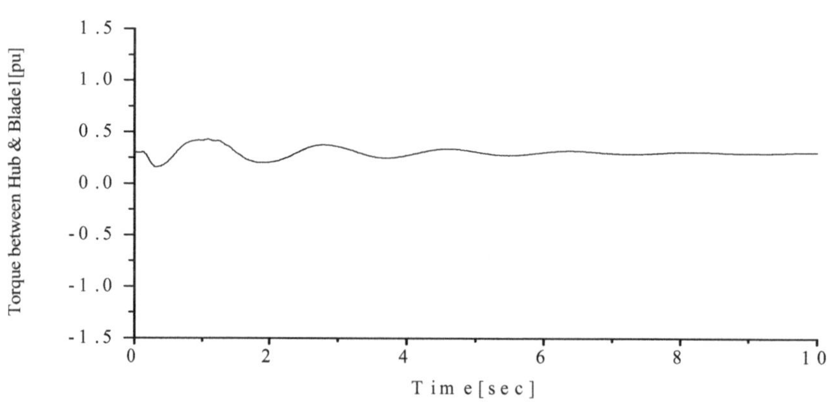

(c) With STATCOM

Fig. 4.39 Torque between the hub and blade of WTGS (symmetrical 3LG fault)

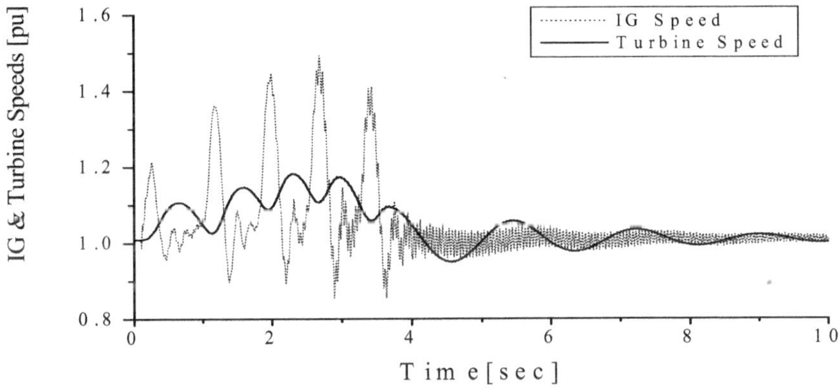

Fig. 4.40 IG rotor and turbine speeds (3LG fault, with pitch controller)

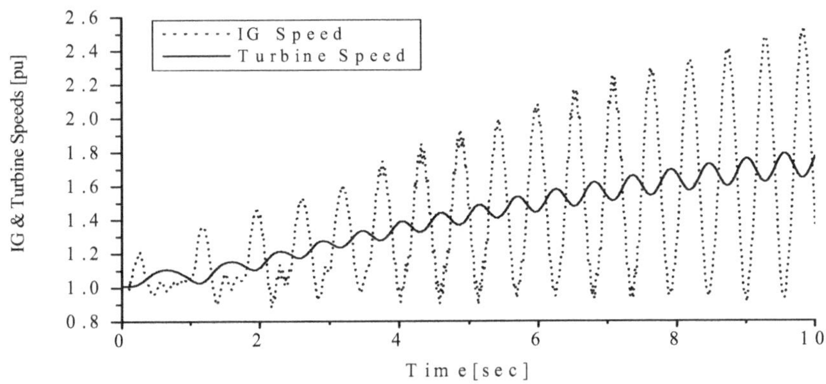

Fig. 4.41 IG rotor and turbine speeds (3LG fault, no controller)

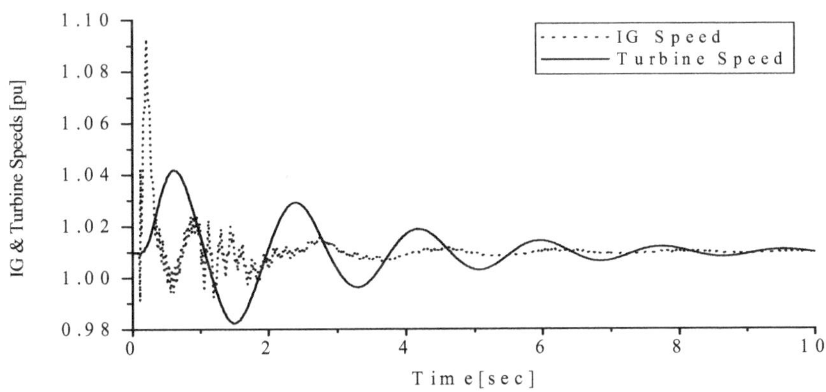

Fig. 4.42 IG rotor and turbine speeds (3LG fault, with STATCOM)

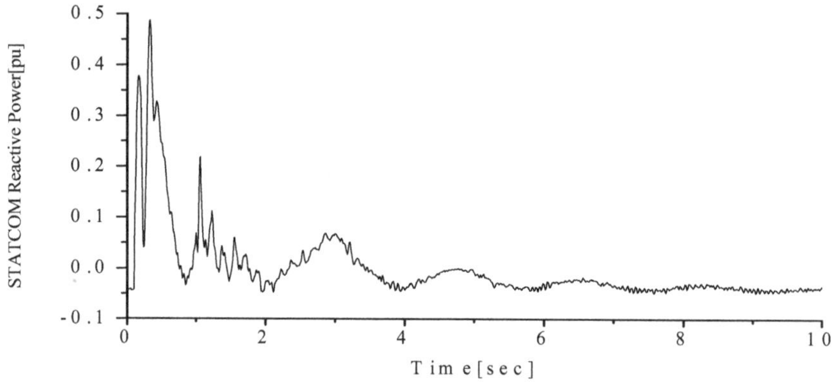

Fig. 4.43 Reactive power of STATCOM (3LG fault)

4.5 Chapter Summary

In this chapter, a three level VSC based STATCOM is proposed to enhance the steady state and transient performance of a fixed speed WTGS connected to a power grid. The detailed modeling and control strategy for STATCOM are presented. It is found that a STATCOM can significantly improve the transient performance of a fixed speed WTGS after a severe network disturbance. It is reported that a STATCOM can reduce the voltage fluctuation of a wind generator significantly under variable wind speed, i.e., it can improve the power quality of a grid connected wind generator.

Another salient feature of this chapter is the analysis of the blade-shaft torsional oscillations of a WTGS during a network disturbance in the power system. It is found that our proposed three-level VSC based STATCOM can significantly decrease the blade-shaft torsional oscillations of a WTGS. Moreover, it is reported that the pitch controller can decrease the blade and shaft torsional oscillations of a WTGS to some degree.

Acknowledgements The authors would like to thank Mr. Masashi Takiguchi for his great help in editing the entire chapter. This help us to finish the book within the specified time limit. Moreover, special thanks to Dr. Mohammad Abdul Mannan for his great help in the simulation analysis.

Chapter 5

Integration of an Energy Storage System into Wind Farm

The oldest form of energy storage involves harvesting ice from lakes and rivers. The ice was stored in well insulated warehouses and sold or used throughout the year for almost everything we use mechanical refrigeration for today, including preserving food, cooling drinks, and air conditioning. The Hungarian Parliament Building in Budapest is still air conditioned with ice harvested from Lake Balaton in the winter.

Chemically charged batteries became quite common in the mid-nineteenth century to provide power for telegraphs, signal lighting, and other electrical apparatus. By the 1890s, central stations were providing both heating and lighting, and many did both. Electric systems were almost all direct current (DC), therefore incorporating batteries was relatively easy. In 1896, Toledo inventor Homer T. Yaryan installed a thermal storage tank at one of his low temperature hot water district heating plants in that city to permit capturing excess heat when electric demand was high. Other plants used steam storage tanks, which were not as successful for some reason.

Other forms of energy storage were used to power streetcars in the 1890s, including compressed air and high temperature hot water that was flashed into steam to run a steam engine. Electric cars and trucks were quite common prior to World War I until gasoline-powered internal combustion engines ran them off the road.

The energy storage system (ESS) is closely associated nowadays with renewable energy sources such as photovoltaic or wind power applications. However, this chapter emphasizes the energy storage systems suitable for wind power application, which will be discussed in the following sections. Much research has been carried out to investigate suitable application of energy storage systems. Sandia National Laboratories played an important role in this arena. This chapter is partly written in the light of one excellent report from that laboratory [132].

5.1 Energy Storage Systems in Power System

5.1.1 Application of Energy Storage Systems

The applications of interest have been classified as bulk energy storage, for load leveling or load management, distributed generation (DG) for peak shaving, and power quality (PQ) or end-use reliability. The different categories are distinguished by the power level and discharge time required. These specifications together determine the stored energy requirement. The power levels and storage times for the various application categories are listed in Table 5.1 [132].

Table 5.1 Application category specifications

Application category	Discharge power range	Discharge time range	Stored energy range	Representative applications
Bulk energy etorage	10 – 1000 MW	1 – 8 hrs	10 – 8000 MWh	Load leveling, spinning reserve
Distributed generation	100 – 2000 kW	0.5 – 4 hrs	50 – 8000 kWh	Peak shaving, transmission deferral
Power quality	0.1 – 2 MW	1 – 30 sec	0.1 – 60 MJ (0.028 – 16.67 kWh)	End-use power quality and reliability

Source: Sandia National Laboratories[1]

The following technologies are available for energy storage systems:

- Lead-acid batteries (flooded and valve-regulated lead-acid, VRLA)
- High temperature sodium/sulfur(Na/S) batteries
- Sodium bromide/sodium polysulfide flow batteries (represented by the Regenesys® system)
- Zinc/bromine (Zn/Br) batteries
- Vanadium-redox (V-redox) batteries
- Lithium-ion batteries (Li-ion)
- Nickel/cadmium (Ni/Cd) batteries
- Superconducting magnetic energy storage (SMES)
- Low-speed flywheels (steel wheel)
- High-speed flywheels (composite wheel)
- Supercapacitors or energy capacitor systems (ECS)

[1] To learn about Sandia National Laboratories, visit at http://www.sandia.gov/

- Compressed air energy storage (CAES) in underground caverns
- Compressed air energy storage in surface vessels (CAES-surface)
- Pumped hydroelectric storage
- Hydrogen storage used with either a hydrogen fuel cell or hydrogen engine

Not all technologies are suitable for all applications, primarily due to limitations in either power output or storage capacity. Table 5.2 below lists the technologies considered for each of the application categories. The third column indicates whether the technology is currently available (A) for this application, or has the potential (P) to be used in this application.

Table 5.2 Technologies considered in each application category

Category	Technology	Available(A) or potential (P)
Bulk energy storage	Lead-acid batteries	A
	Na/S batteries	P
	Regenesys	A
	Zn/Br batteries	A
	Ni/Cd	A
	CAES	A
	Pumped hydro	A
Distributed generation	Lead-acid batteries	A
	Na/S batteries	A
	Ni/Cd	A
	Li-ion batteries	A
	Zn/Br batteries	A
	V-redox batteries	A
	High-speed flywheels	P
	CAES-surface	P
	Hydrogen fuel cell	A
	Hydrogen engine	A
Power quality	Lead-acid batteries	A
	Li-ion batteries	P
	High-speed flywheels	A
	Low-speed flywheels	A
	SMES	A
	Supercapacitors	A

Source: Sandia National Laboratories

5.1.2 System Description

For the three application categories, a system configuration was assumed so that system costs and performance could be estimated. The bulk storage and distributed generation systems are shown schematically in Fig. 5.1. The DC storage unit is assumed to interface the AC electric grid through a power conversion system (PCS) that operates only when dispatched as a source or load. The PCS is rated at the power level (kW, MW) required for the application, and the energy storage unit is rated (kWh, MWh, MJ) to provide power for the required duration. For some technologies, the energy storage unit may be oversized if it cannot be completely discharged in a short time [132].

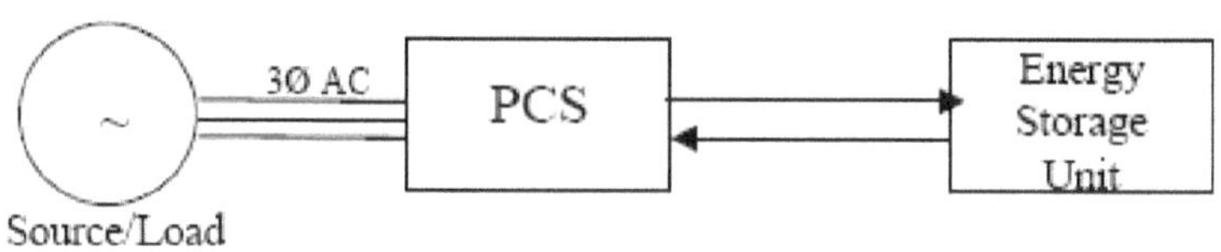

Fig. 5.1 An energy storage system connected directly to an electric grid via a power conversion system (Source: Sandia National Laboratories)

In a power quality or end-use application, the energy storage system may be connected to the bus that feeds a user's load such as a machine or industrial processing unit. In this case, the storage unit is activated only when the grid power is disrupted, but it must be in communication with the bus at all times so that operation is nearly instantaneous whenever a disturbance occurs in the system. This configuration is shown schematically in Fig. 5.2. It can be implemented in several ways. In one implementation, the PCS is continuously energized and the energy storage unit may be trickle charged, resulting in energy losses due to PCS inefficiencies and storage unit charging. In another implementation of this configuration, the system may include a fast, high power switch that can connect the PCS and storage unit to the bus in about 4 milliseconds, which is a seamless connection for almost all loads. This implementation incurs fewer energy losses during normal operation, but requires the installation and maintenance of a fast switch.

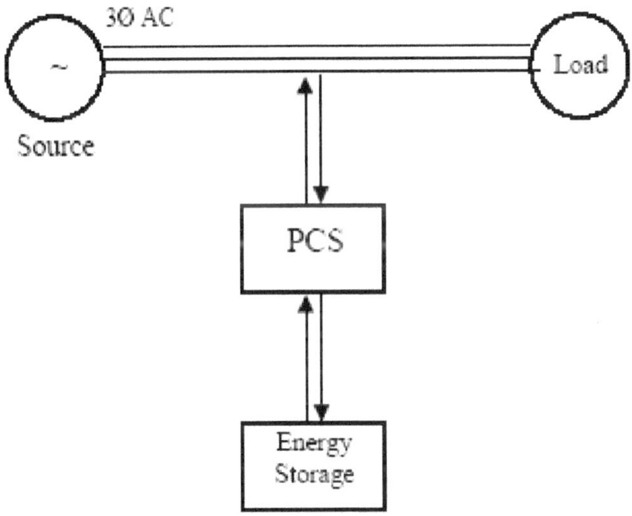

Fig. 5.2 An energy storage system connected to a bus that feeds the load (Source: Sandia National Laboratories)

5.2 Use of Power Electronics in ESS

Power electronics, in general terms, is defined as the use of switching devices to control and convert electrical power flow from one form to another to meet a user's need. "Convert" is a general term used in power electronics to describe the process of changing power from one form to another. The hardware that performs the process is generally called the converter. Converters can perform the following processes/conversions (when each process/conversion is performed, the hardware is referred to by a particular name):

Table 5.3 Electrical power conversion

Conversion	Common Names
AC-to-DC	Rectifier
DC-to-AC*	Inverter
DC-to-DC	boost, buck, buck-boost, chopper, etc.
AC-to-AC	cycloconverters

*The most common type of conversion is DC-to-AC (inversion); e.g., converting DC power from a storage device into AC power for use by a utility grid or other end-user.

Performing the conversions requires some essential hardware: a control system, semiconductor switches, thermal management devices, protection devices, magnetics such as transformers and filters, DC and AC disconnects, and enclosures. Taken together, this hardware is referred to as the power conversion system (PCS). Among different types of semiconductor switches, diodes, gate turn-off thyristor (GTO), and insulated-gate-bipolar transistor (IGBT) are frequently used in a PCS.

The rest of the sections of this chapter emphasize the energy storage system (ESS) for power quality (PQ) issue because it is suitable for wind power application.

5.3 Energy Storage System for Wind Power Application

In the introduction, it is reported that 94.1 GW of wind power installation capacity was achieved at the end of 2007. It is expected that this growing trend will remain and that wind power will play a very important role in the near future.

However, one of the largest disadvantages of wind power production is its strong dependence on the weather. Due to sudden and large changes in wind speed, the power output from a wind farm can have large fluctuations.

Between the two types (fixed and variable speed) of wind turbine generator systems, the fixed speed wind generator also has terminal voltage fluctuation due to the randomly fluctuating wind speed. The variable speed wind generator is connected to the network through a fully controlled frequency converter. Therefore, terminal voltage fluctuation is not present in the variable speed wind generator. However, both types of wind turbine generator systems have output power fluctuation.

An energy storage system can be an effective tool for wind power application that can help a wind farm to stabilize its power and voltage fluctuating tendency. The energy storage system has both real and reactive power control ability. An ESS can be connected at a wind farm terminal. Therefore, both fixed and variable speed wind farm output power fluctuation can be smoothed by providing or absorbing the necessary real power from or to an ESS. Additionally, the fixed speed wind farm terminal voltage fluctuation can also be smoothed by controlling the reactive power when ESS is connected at a wind farm terminal. Figures 5.3 and 5.4 show the schematic diagram when an ESS is adopted by fixed and variable speed wind farms, respectively.

In Chap. 7 of this book, more detail will be discussed for a particular type of energy storage system. Note that the size of the ESS for a wind power application depends particularly on the location of the wind farm. For example, in Japan, the large wind speed fluctuation demands a comparatively larger ESS than that in Europe.

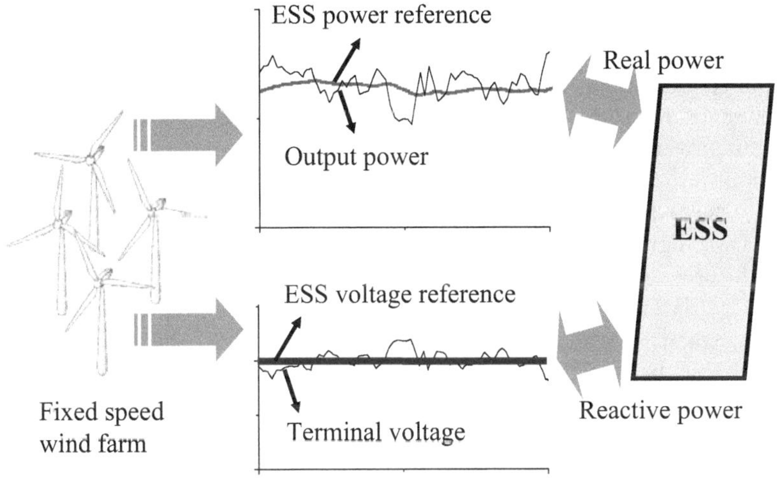

Fig. 5.3 Schematic diagram of integrating an ESS with a fixed speed wind farm

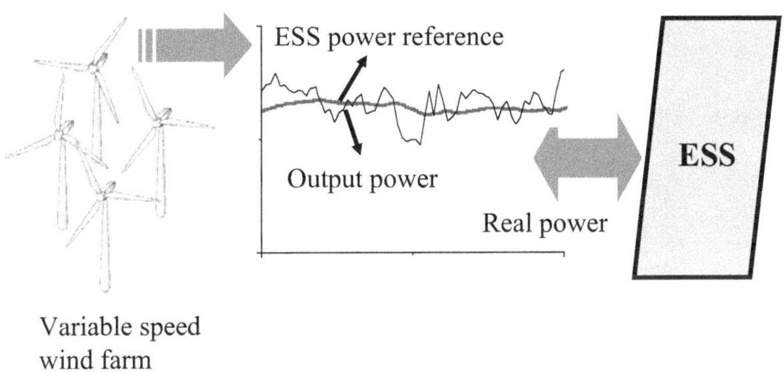

Fig. 5.4 Schematic diagram of integrating an ESS with a variable speed wind farm

The following four types of energy storage systems can be adopted suitably by wind farm:

- STATCOM integrated with battery energy storage system (STATCOM/BESS)
- Flywheel energy storage system (FESS)
- Superconducting magnetic energy storage (SMES) system
- Energy capacitor system (ECS).

5.3.1 STATCOM/BESS

In Chap. 4, it has been shown that a STATCOM can be a very effective tool for stabilizing a fixed speed wind farm, especially for transient performance, voltage quality improvement, and to reduce the blade-shaft torsional oscillation during a network disturbance. However, a STATCOM has only the reactive power controllability. But when an energy storage system (ESS) is integrated with a STATCOM, it gives excellent controllability of both real and reactive power. This section focuses on a STATCOM with battery energy storage system (BESS), i.e., STATCOM/BESS topology for a wind power application.

The schematic diagram of STATCOM/BESS topology is shown in Fig. 5.5. A two-level VSC based STATCOM shown inside the dotted lines in Fig. 5.5, controls only the reactive power output. Therefore, BESS is incorporated with a S-TATCOM, resulting in control of both real and reactive power. In the traditional STATCOM, the DC-link capacitor is extremely large, whereas in the STATCOM/BESS topology, a small capacitor can be used to smooth the DC current [79].

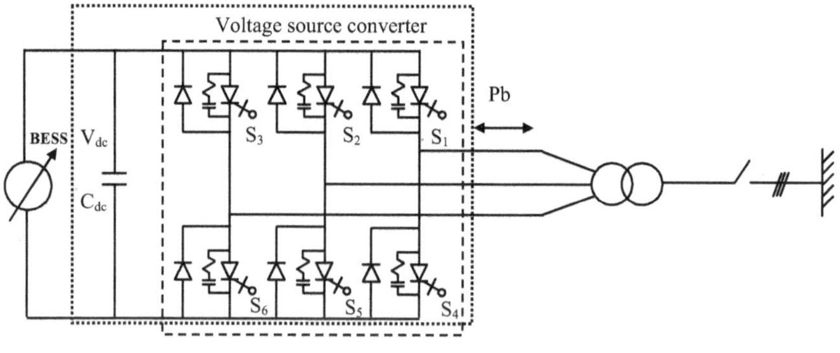

Fig. 5.5 Schematic diagram of STATCOM/BESS topology

The VSC of a STATCOM can be a two-level or multi-level structure, depending on the operating voltage. A STATCOM/BESS can be connected to a grid in the same way as that shown in Fig. 5.2. Switching devices such as GTOs and IGBTs are suitable for a PCS of a STATCOM/BESS. Because of the detailed discussion of the STATCOM in Chap. 4, only discussion of a BESS is carried out in the rest of this section.

5.3.1.1 Battery Classification

A battery can be classified in different ways. However, generally, a battery is classified as follows [132]

5.3.1.1.1 Primary Battery

A primary battery is not rechargeable. It can be operated until the reactants are consumed. It must then be discarded in an environmentally safe manner.

5.3.1.1.2 Secondary Battery

A rechargeable battery is also called a secondary battery. Applying an electric current or voltage can reverse the direction of the reactions and thus recharge the battery. Energy is again stored in the chemical potential of the reactants. This is the most common type of rechargeable battery. Most consumer products with rechargeable batteries, such as laptop computers, cameras, and remote control toys, use conventional rechargeable batteries. Most are either lead-acid, nickel-metal hydride, nickel/cadmium, or lithium-ion batteries. In conventional rechargeable batteries, the unit is self-contained and the only thing that flows in and out of the battery is electricity. A generic rechargeable battery is shown in Fig. 5.6. In the charging mode, electricity is applied to the battery and ions flow in the opposite direction.

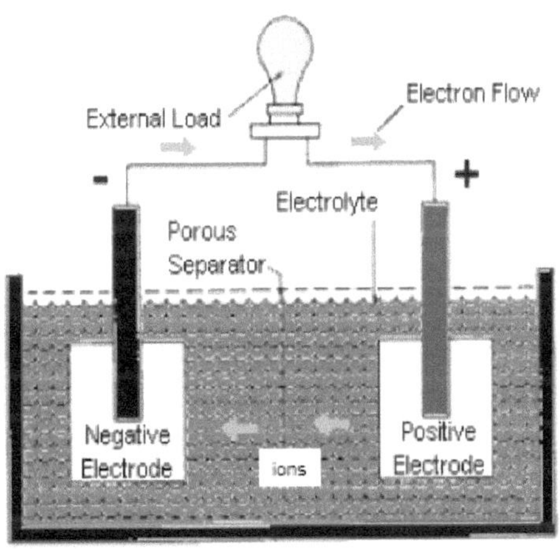

Fig. 5.6 Generic rechargeable battery discharging

5.3.1.1.3 Flow Batteries

Flowing electrolyte batteries generally contain pumps, plumbing, electrolyte reservoirs, and electrochemical cell stacks. During charge and discharge operations, the battery electrolyte is circulated through the cell stacks in which the electrochemical reactions take place. Charged species may be stored inside the stacks or in the reservoirs. Redox flow batteries are "those electrochemical systems where the oxidation and reduction of two chemical species take place on inert electrodes and these active materials are stored externally from the battery cell" [133]. In operation, the reactants flow through opposite sides of a cell, separated by an inert separator. In such a system, the storage capacity is determined by the mass of reactants (as in all batteries), but the capacity is easily increased or decreased by changing tank sizes. Flow batteries are rechargeable; the reactants are good for a minimum of 2000 cycles, and up to more than 10,000 cycles [134]. The overall battery system is also self-contained, i.e., from the users' point of view the only thing that flows in and out of the system is electricity. Three common types of flow batteries are zinc/bromine batteries, vanadium-redox batteries, and the system called Regenesys®. A vanadium-redox flow battery is shown in Fig. 5.7. In this battery, the two active species are vanadium in different states of oxidation. Another rechargeable flow battery is the Regenesys® system produced by Innogy, PLC, in the UK. This system uses polysulfide as the active species.

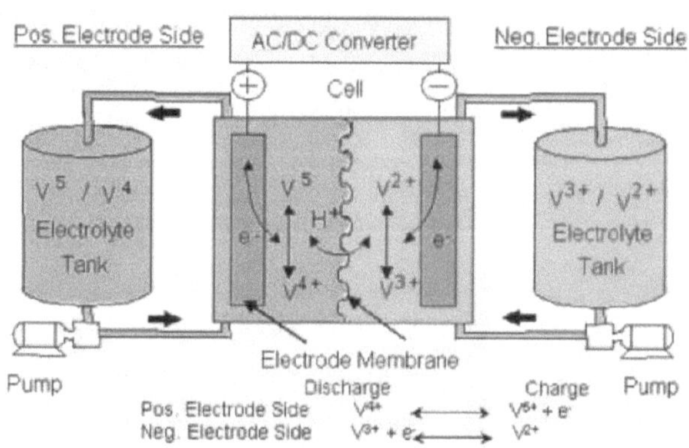

Fig. 5.7 Vanadium-redox battery

5.3.1.2 Batteries for Power Quality (PQ) Systems

In general, for power quality (PQ) systems, discharge is limited to less than one minute. Storage is assumed around 10 minutes. From the above-mentioned classification of batteries, only a few may be suitable for power quality systems. The lead-acid battery is frequently used for PQ systems. The lead-acid battery was successfully used with many systems and still enjoys big demand because it is a mature technology. Sandia LAB has a wide experience (over 60 MW in the field) and cost information on lead-acid battery. Individual modules are rated at 250 kW. Discharge is limited to short duration, up to 30 seconds. Longer discharge is possible under a variety of conditions. The batteries, however, store about 40 kWh at the 10-minute discharge rate. The battery module costs $12K for a 250-kW system, or $300/kWh. This module must be replaced every six years. Cost of the PCS is $410/kW at this size and will decrease to about $250/kW for larger sizes. There is no clear need for O&M, fixed or variable. The Li-ion battery can also be used for PQ systems. However, like most batteries, Li-ion batteries cannot be completely discharged in a few seconds. Like a PQ lead-acid battery, a 10-minute storage system is assumed. Another difference is in the way the converter losses are handled. Because the converter may be connected full time, there can be continuous converter losses. The vanadium-redox batteries are a relatively new energy storage technology, usually used in distributed generator systems but can also be used in PQ systems. It can also be integrated at the wind farm terminal for minimizing the voltage and power fluctuation due to randomly fluctuating wind speed.

A 170 kW-6hr capacity vanadium-redox battery system is already installed beside an experimental wind power generator rated at 275 kW in a plant of Hokkaido Electric Power Company located at Tomari Hill. The experiment is conducted jointly by NEDO and the Institute of Applied Energy to assess the system's effectiveness in smoothing electricity output.

5.3.2 Superconducting Magnetic Energy Storage (SMES) System

Basic discussions on a SMES unit and a topological overview are presented in this section.

5.3.2.1 Basic Discussion and Applications of SMES

Superconducting magnetic energy storage (SMES) systems store energy in the magnetic field created by the flow of direct current in a superconducting coil that has been cryogenically cooled to a temperature below the superconducting critical temperature. The SMES system includes mainly three parts: a superconducting coil, a power conditioning system, and a cryogenically cooled refrigerator. Once the superconducting coil is charged, the current will not decay and the magnetic energy can be stored indefinitely. The stored energy can be released back to the network by discharging the coil. The power conversion system (PCS) uses an inverter/converter to transform AC power to direct current or convert DC back to AC power. The inverter/rectifier accounts for some energy loss in each direction. The SMES loses the least amount of energy in storage process compared to other methods of storing energy. SMES systems are highly efficient; the round trip efficiency is greater than 95%. Due to the energy requirements of refrigeration and the high cost of superconducting wire, the SMES system is currently used for short duration energy storage. Therefore, a SMES system is most commonly devoted to improving power quality. If a SMES system were to be used for utilities, it would be a diurnal storage device, charged from base-load power plants at night and meeting peak.

A superconducting magnetic energy storage (SMES) system, designed to improve the power quality for critical loads, provides carryover energy during voltage sags and momentary power outages. The system stores energy in a superconducting coil immersed in liquid helium. Figure 5.8 shows a basic schematic diagram of a SMES system.

Utility system power feeds the power switching and conditioning equipment that provides energy to charge the coil, thus storing energy. When a voltage sag or momentary power outage occurs, the coil discharges through switching and conditioning equipment, feeding conditioned power to the load.

The refrigeration system and helium vessel keep the conductor cold to maintain the coil in the superconducting state. The user can place the magnetic coil either in parallel (shunt-connected) or in series (series-connected or in-line) with the power conditioning equipment.

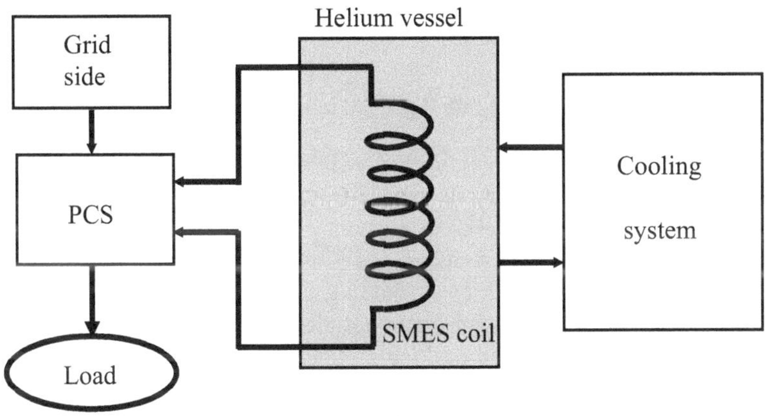

Fig. 5.8 Schematic diagram of a basic SMES system

5.3.2.2 Modeling and General Control Strategy

The state variable equations defining the voltage and current deviations in the inductor can be expressed by Eqs. 5.1 and 5.2:

$$\frac{d}{dt}\Delta V_{sm} = \frac{1}{T_{dc}}\left(K_0 \Delta f - \Delta V_{sm}\right) \tag{5.1}$$

$$\frac{d}{dt}\Delta I_{sm} = \frac{1}{L}\left(\Delta V_{sm} - R_L \Delta I_{sm}\right) \tag{5.2}$$

For simplicity in comprehending the physical extent of the excursions of the variables belonging to a SMES unit, ΔV_{sm} and ΔI_{sm} have been expressed in kV and kA, respectively. R_L and L are expressed in Ω and H, respectively, and K_0 is expressed in kV/Hz.

The current and voltage of a superconducting inductor are related by

$$I_{sm} = \frac{1}{L_{sm}} \int_{t_0}^{t} V_{sm} d\tau + I_{sm0} \tag{5.3}$$

where, I_{sm0} is the initial current in the inductor.

The energy stored in the superconducting inductor can be expressed by the following equation:

$$W_{sm} = W_{sm0} + \int_{t_0}^{t} P_{sm} d\tau \tag{5.4}$$

where, $W_{sm0} = \frac{1}{2} L_{sm} I_{sm}^2$ is the initial energy in the inductor.

Energy storage in a power system can reduce the time or rate mismatch between energy supply and energy demand, thereby playing a vital role in energy conservation. Energy storage leads to a saving of premium fuels and makes the system more cost-effective by reducing the wastage of energy. It improves the performance of energy systems by smoothing and increases the reliability. Because of this, energy storage is an important element in many utility systems. It provides a means for easing load peaking problems and improving the load factor in base-load plants.

A SMES system is inherently very efficient and has sitting requirements that are different from those of other technologies. Because of these characteristics, a SMES system has the potential of finding application in systems with large energy storage requirements. This recently conceived technology meets many of the utility's requirements for diurnal storage. A usual feature of a SMES unit is the cost scaling with size, which is different from that of other storage devices. For a given design, the cost of a SMES unit is roughly proportional to its surface area and the required quantity of superconductor. The cost per unit of stored energy (megajoules or kilowatt-hours) decreases as storage capacity increases.

In addition, the charge and discharge of a SMES unit is through the same device, a multiphase converter, which allows the SMES system to respond within tens of milliseconds to power demands that could include a change from maximum charge rate to maximum discharge power. This rapid response allows a diurnal storage unit to provide a spinning reserve and to improve system stability. Both the converter and the energy storage in the coil are highly efficient because there is no conversion of energy from one form to another as in pumped hydro, for example, where the electrical energy is converted to mechanical energy and then back again. The major loss during the storage is the energy required to operate the refrigerator that maintains the superconducting coil in a superconducting state. Because of these characteristics and because it can be easily sited, a SMES has the potential of finding extensive application in electric utility systems.

Again in power systems, continuous reactive compensation of the load end of the transmission lines is generally required for static and dynamic voltage control and preservation of the system stability. The active and reactive power of a SMES system can be controlled easily. In that case, the SMES system can be independently controlled to give freedom in selecting the VAR consumption at any point in the active power transfer.

Figure 5.9 shows the basic configuration of a SMES unit in the power system. The voltage source converter (VSC) consists of a PWM rectifier/inverter using IGBTs or GTOs. The control concept of a SMES system charging and discharging energy is shown in Fig. 5.10. The DC-DC chopper is controlled to supply positive (IGBT is turned on) or negative (IGBT is turned off) voltage V_{sm} to the SMES coil, and then the stored energy can be charged or discharged. Therefore, the superconducting coil is charged or discharged by adjusting the average voltage, V_{sm-av}, across the coil which is determined by the duty cycle of the two-quadrant DC-DC chopper.

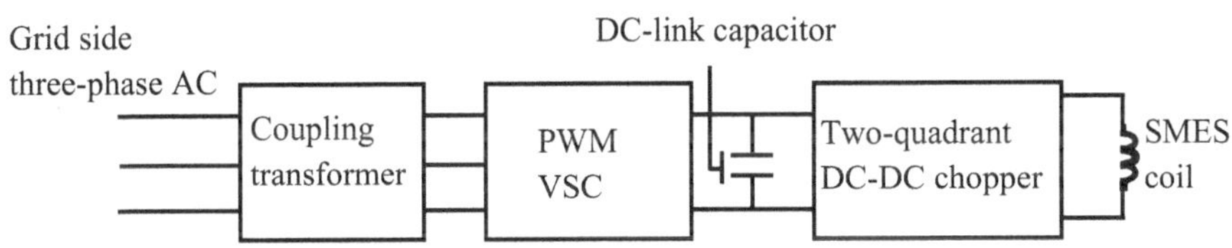

Fig. 5.9 Basic components of a SMES control system

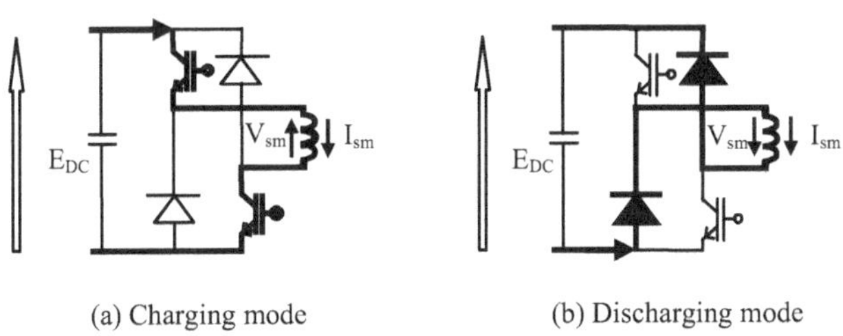

Fig. 5.10 Control concept of energy charging and discharging in SMES

A SMES system can be used effectively for wind farm output power smoothing. In some recent studies, a SMES system is proposed for integration into the wind farm [73 – 75].

5.3.3 Flywheel Energy Storage System (FESS)

5.3.3.1 Working Basics

Flywheels are kinetic energy storage systems. Using a simplified example, a flywheel's function is the same as that of batteries. That means, energy storage, but the technology used for energy storage is different.

Batteries, in general, use electrochemical processes that allow battery recharging and energy storage. Flywheels transform electrical energy into kinetic energy and the other way round, so that the energy is stored as kinetic energy.

Basically, a flywheel is composed of a shaft that integrates a flywheel. A rotor of an electrical machine is mounted on that shaft. The flywheel housing contains the electrical machine stator and other elements needed for the appropriate functioning of the machine, for example, the shaft bearings. The flywheel also includes electronics, which is not explained in this chapter.

When electrical energy has to be transformed into kinetic energy, the electrical machine works as a motor that absorbs electrical energy accelerating the shaft until the working speed is reached.

Once that speed has been reached, the electrical machine is disconnected from the net, but the shaft, due to the inertia of the flywheel, goes on rotating for a very long time. In this way, electrical energy has been transformed into kinetic energy, and therefore, energy is stored as the shaft is rotating.

To get the shaft rotating indefinitely, mechanical energy losses, such us, bearing friction, aerodynamic losses, and so on must be eliminated. For that purpose, different solutions have been adopted. For example, a vacuum atmosphere or helium atmosphere is created inside the housing.

When the stored energy must be extracted from the machine, the kinetic energy is transformed into electrical energy. In this case, the electrical machine works as a generator, whose shaft is already in movement. In this way, electrical energy is obtained through the generator, as the shaft speed is reduced.

The electrical parameters of the extracted electrical energy are controlled by the electronics to have the appropriate tension, frequency, and power.

5.3.3.2 Material Used in Flywheel

There are two basic classes of flywheels based on the material used in the rotor. The first class uses a rotor made up of an advanced composite material such as carbon-fiber or graphite. These materials have very high strength to weight ratios, which give flywheels the potential of having high specific energy. The second class of flywheels uses steel as the main structural material of the rotor. This class includes traditional flywheel designs that have large diameters, rotate slowly, and have low power and energy densities, but also includes some newer high performance flywheels as well. In an integrated flywheel, the energy storage accumulator and the electromagnetic rotor are combined in a single-piece solid steel rotor. By using an integrated design, the energy storage density of a high power steel rotor flywheel energy storage system can approach that of a composite rotor system, but avoids the cost and technical difficulties of a composite rotor.

5.3.3.3 Development/Deployment Status

While high-power flywheels are developed and deployed for aerospace and uninterruptible power supply (UPS) applications, there is an effort, pioneered by Beacon Power, to optimize low cost commercial flywheel designs for long duration operation (up to several hours). 2 kW/6 kWh systems are used in telecom service today[2]. Megawatts for minutes or hours can be stored using a flywheel farm approach. Forty 25 kW/25 kWh wheels can store 1 MW for 1 hour efficiently in a small footprint. A Beacon 25 kWh flywheel is shown in Fig. 5.11.

The stored energy can be approximated by

$$E = \frac{I\omega^2}{2} = \frac{mr^2\omega^2}{2} = \frac{mv^2}{2} \tag{5.5}$$

Where, ω is the rotational velocity (rad/sec), I the moment of inertia for the thin rim cylinder, m is the cylinder mass, and v is the linear rim velocity.

Flywheel energy storage systems are widely used in space, hybrid vehicles, the military field, and power quality. Space station, satellites, and aircraft are the main applications in space. In these fields, flywheel systems function as energy storage and attitude control. For the applications in hybrid vehicles and the military field, flywheel systems are mostly used to provide pulse power. But for power quality application, flywheel systems are widely used in USP, to offer uninterruptible power and voltage control.

A flywheel energy storage system (FESS) can be used to stabilize a wind farm effectively. Several studies are reported in this area, where a flywheel is proposed for integration with the wind generator [66 – 68]. Figure 5.12 shows a schematic diagram that consists of a wind farm, a flywheel, a consumer load, and the main power plant.

---

[2] To find the details of Beacon Products, visit at http://www.beaconpower.com/

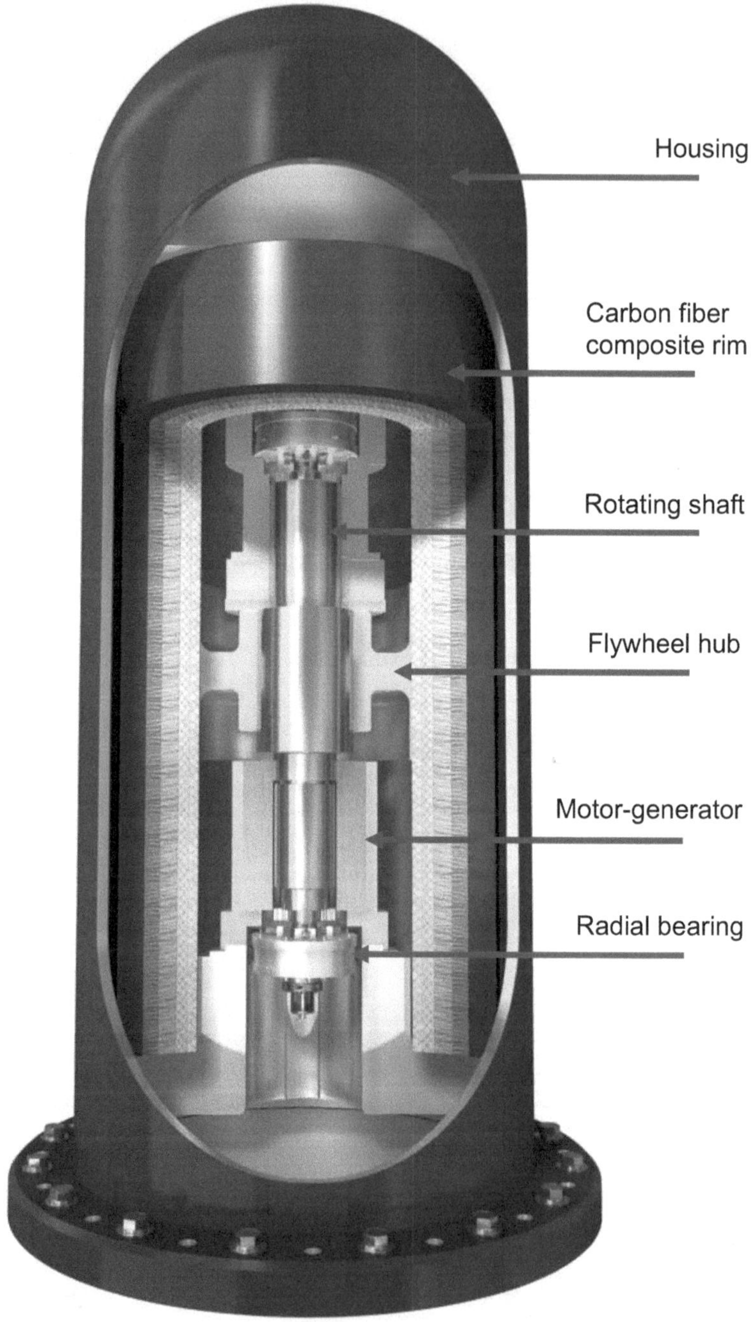

Fig. 5.11 Beacon Power Smart Energy 25 flywheel (Courtesy Beacon Power Corporation, US)

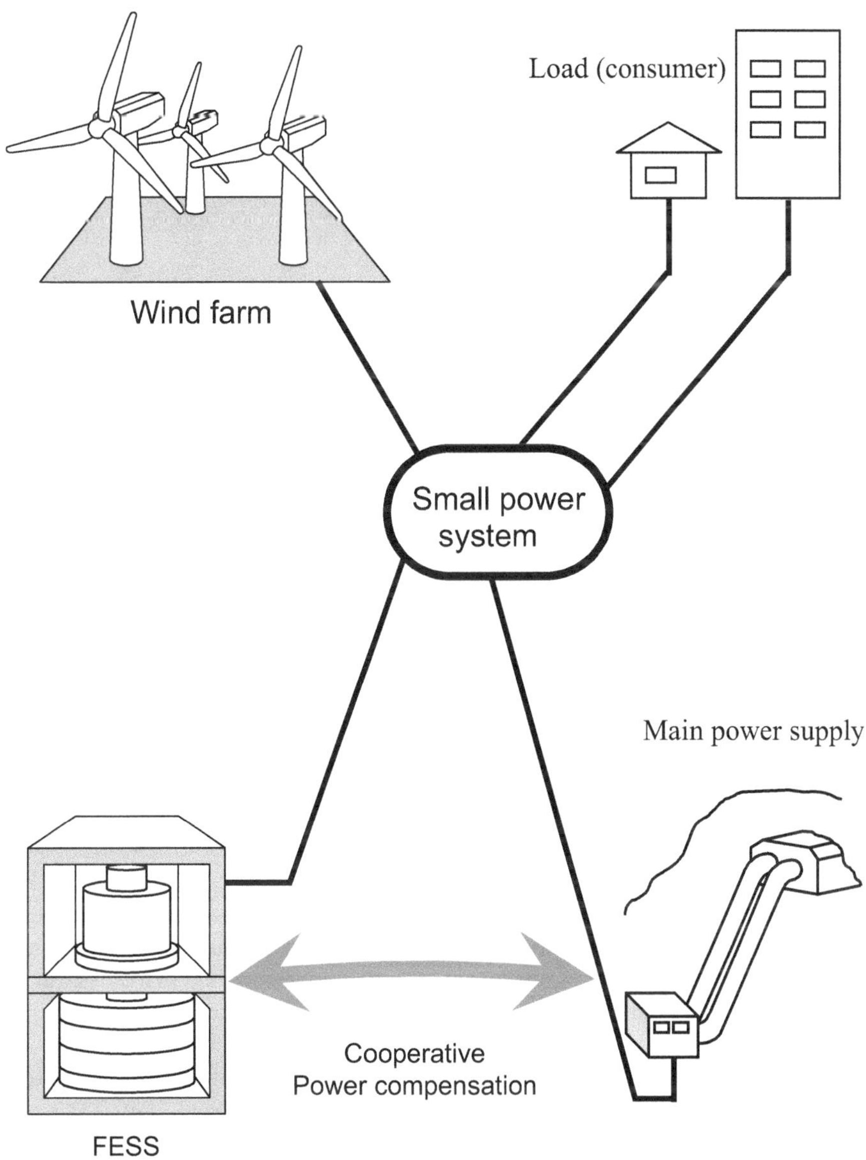

Fig. 5.12 Schematic diagram of a FESS application to a wind farm

5.3.4 Energy Capacitor System (ECS)

An energy capacitor system (ECS) consists of an EDLC and power electronic devices and is used as an energy storage system. The basic components of an ECS control system are shown in Fig. 5.13, which composed of a PWM voltage source converter (VSC), a DC-DC buck/boost converter, and an EDLC bank. The PWM VSC controls the DC-link voltage and reactive power flowing from the grid, whereas the DC-DC buck/boost converter controls the real power.

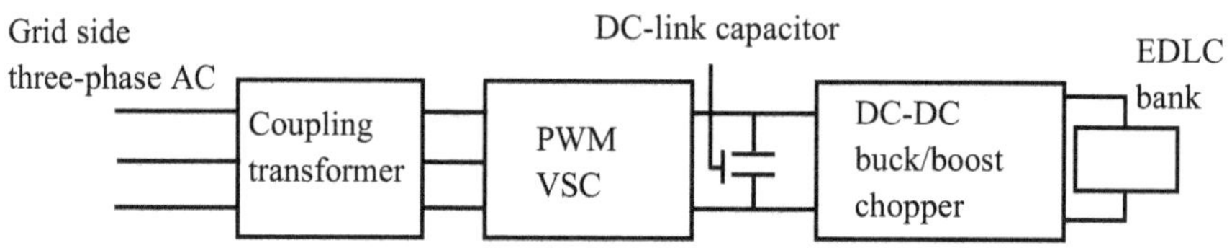

Fig. 5.13 Basic components of an ECS control system

5.3.4.1 Theory and Modeling of EDLC

5.3.4.1.1 EDLC Overview

During the past few years, a pollution-free high-performance electric energy storage device has been demanded. One of the best solutions is the electric or electrochemical double layer capacitor (EDLC), also known as the super capacitor or the ultra capacitor. Professor Pieter van Musschenbroek invented the first capacitor in 1745 by accident. Later, in the middle of the nineteenth century, the German physicist Herman Ludvig Ferdinand von Hemholtz formulated the principle of an electrochemical double layer capacitor. It took almost a century before some scientists started to develop Hemholtz's idea, and at the beginning of 1980, the Japanese were the first to succeed in the technical realization, when a bulky sized capacitor with a capacitance of 10 farads came onto the market. After that, scientists and manufacturers invested much time in this technology, which has become very widespread in the whole electric world.

5.3.4.1.2 EDLC Principle

A basic capacitor consists of conductive foils and a dry separator. There are three types of electrode materials adequate for an EDLC. One of the most common is high surface area activated carbon. It is also the cheapest to manufacture. The two other electrode materials are metal oxide and conducting polymers, but they are not as commonly used as activated carbon.

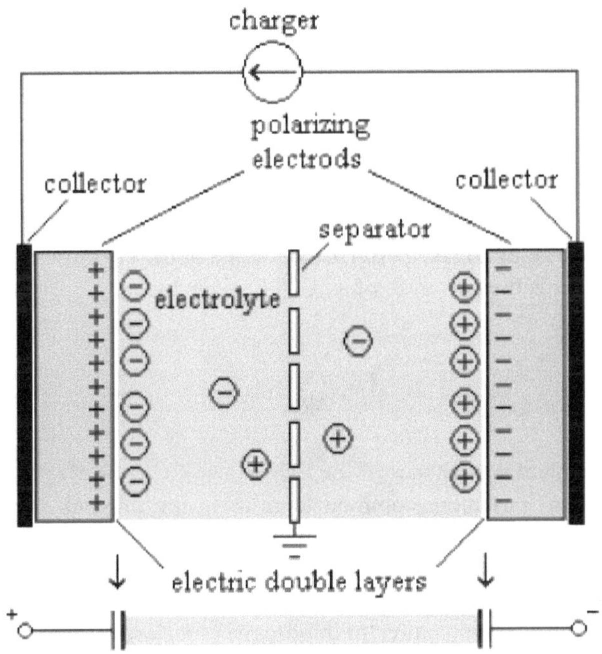

Fig. 5.14 Principle of the electric double layer capacitor [134]

The EDLC is a charge storage device, which utilizes a double layer formed on a large surface area of a micro porous material such as activated carbon. The structure of the double layer is shown in Fig. 5.14. The EDLC stores the energy in the double layer formed near the carbon electrode surface. There are two layers: one layer of electrolyte molecules and a second layer for diffusion. In the first layer, the electrons cannot move at all. In the second, layer the electrons can move around a little [135].

5.3.4.1.3 Principle of Energy Storage

An EDLC stores electric energy in an electrochemical double layer formed at a solid/electrolyte interface. Positive and negative ionic charges within the electrolyte accumulate at the surface of the solid electrode and compensate for the electronic charge at the electrode surface.

The energy storage capacity of an EDLC can be described by Eq. 5.6:

$$E = \frac{1}{2}CV^2 \tag{5.6}$$

where, E is the stored energy in joules (J), V is the rated or operating voltage of the EDLC, and C is the capacitance (F).

Apart from the voltage limitation, the size of the EDLC controls the amount of energy stored, and the distinguishing feature of an EDLC is its particularly high capacitance.

Another measure of EDLC performance is the ability to store and release energy rapidly. This is the power, P, of an EDLC given by

$$P = \frac{V^2}{4R} \tag{5.7}$$

where, R is the internal resistance of the EDLC

For capacitors, it is more common to refer to the internal resistance as the equivalent series resistance (ESR). The power performance is controlled by the ESR of the entire device, and this is the sum of the resistance of all the materials, that is, substrate, carbon, binder, separator and electrolyte, between the external contacts. Therefore, optimization of high-power performance can be achieved through an understanding of the nanostructure of the materials and the nanoprocesses that dictate the EDLC performance.

5.3.4.1.4 Advantages and Disadvantages

The EDLC is a widely broadening component in the whole power electronic field especially nowadays when simple and workable methods are highly recognized. The EDLC has many good features. Virtually unlimited cycle life is the best among them. The EDLC can be fully charged in seconds, and it can be cycled millions of time.

The EDLC has a simple charging method; there is no need to build any protective circuits. Overcharging or overdischarging does not have a negative effect on the lifespan, as it does on that of chemical batteries. After a full charge, it stops accepting charge.

EDLCs do not harm the environment because they do not contain pollutants like some batteries. Most batteries contain toxic materials. Ni-Cd batteries contain cadmium (Cd) and lead-acid batteries contain lead (Pb). The EDLC doesn't contain heavy metals or toxic materials like Ni, Cd, Pb. There are three types of electrode materials suitable for an EDLC as explained in Sect. 5.3.4.1.2. Therefore, an EDLC is more environment friendly than batteries.

The electric double layer capacitor is not an ideal component. There are some limitations. The cells have low voltage, and if there is a need for a higher voltage, a series connection is needed. If there are more than three capacitors in series, voltage balancing is required. It will extend the board space radically, especially in portable applications and that could be crucial for the whole system. EDLCs have a high self-discharge rate. After one month, the charge of the capacitor decreases from full to 50 %. On the other hand, the EDLC is a longlasting capacitor; it deteriorates to 80 % after one decade in normal use. Also, one of the disadvantages is its low energy density. EDLCs usually hold one-fifth to one-tenth of the energy density of an electrochemical battery [136].

5.3.4.1.5 Examples of Application

Applications of double layer capacitors have grown enormously during the last few years and manufacturers are currently developing them to get optimal features in reasonable packages. Electrical double layer capacitors are used in applications, such as the following [135,137]:

- Backup power sources for portable application in power failure
- Memory backup for programs, timers, etc.
- Power sources for equipment that uses photovoltaic cells
- Underground networks (subway systems) support
- Uninterruptible power supply (UPS). It supplies, e.g., emergency generators
- Starter for small motors
- Power system oscillations damping
- Wind farm output power smoothing and terminal voltage regulation.

5.3.4.1.6 EDLC Modeling

There are a few types of EDLC models available for simulation study. One is the simplified equivalent lumped model [83] of the EDLC cell that can be expressed as shown in Fig. 5.15a. In some literature [87, 89, 90], the distributed model shown in Fig. 5.15b is also considered to represent the terminal characteristics of the EDLC cell precisely, where the distributed parameters are determined by certain percentages of the lumped model.

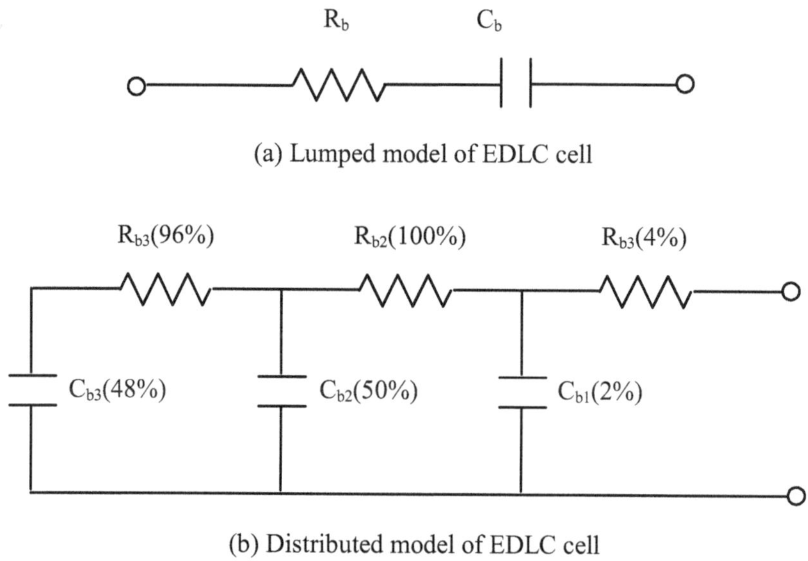

Fig. 5.15 Equivalent circuit of an EDLC cell

An energy capacitor system can also be a good choice for wind power application due to some advantages over other energy storage systems. In some recent papers, it is shown that an ECS can be successfully used to smooth wind farm output power and terminal voltage fluctuation [90, 114]. In Chap. 7, the application of an ECS for wind farm output power and terminal voltage smoothing is presented in detail. Moreover, it is shown that an ECS can enhance the low voltage ride through (LVRT) capability of a wind farm.

5.4 Cost/Performance Analysis

For the above mentioned four types of technologies, Sandia National Laboratories presented a nice cost/performance analysis for power quality (PQ) shown in Table 5.4 [132].

Table 5.4 Cost/Performance analysis for power quality issue (Source: Sandia National Laboratories)

Technology	Energy-related cost (delivered) ($/kWh)	Power-related cost ($/kW)	Efficiency (AC-AC)	Replacement cost ($/kWh)	Replacement frequency (yr)	Parasitic loss converter (kW/kW)	Parasitic loss storage (kW/kW)	Fixed O&M ($/kW-yr)
Lead-acid batteries	300	250	0.75	300	6	0.002	0.00001	10
Li-ion	500	200	0.85	500	10	0.002	0.0001	10
Micro-SMES	50,000	200	0.95	0	None	0.002	0.01	10
Flywheels (high-speed) 150 kW for 15 min.	1,000	300	0.95	0	None	0.002	0.0005	5
Flywheels (high-speed) 120 kW for 20 sec.	24,000	333	0.95	16,000	16	0.002	0.0005	5
Flywheels (high-speed) 200 kW for 20 sec.	125,000	300	0.95	0	None	0.002	0.002	5
Flywheels (low-speed)	50,000	300	0.9	0	None	0.002	0.002	5
Supercapacitors	30,000	300	0.95	0	None	0.002	0.0001	5

Note: The kW parameters in the first column are included only for FES because the costs are specific to those systems and are not generic

5.5 Chapter Summary

In this chapter, different types of energy storages systems such as STATCOM/BESS, SMES, flywheel, and ECS are discussed in detail, including the basic configuration, connection scheme, and application. Because a wind farm is strongly dependent on the weather, it is necessary to incorporate some energy storage system (ESS) at the wind farm terminal to smooth its output power and terminal voltage. Therefore, in this chapter, four types of energy storage systems suitable for wind power application are emphasized. From the four types of energy storage systems, the use of an energy capacitor system (ECS) for wind power will be discussed in detail in Chap. 7 of this book.

Acknowledgments Special thanks to Dr. Rion Takahashi for preparing an illustration used in this chapter. The authors thank Mr. M. R. I Sheikh for providing the necessary help while writing the section of SMES.

Chapter 6

Hydrogen Generation from Wind Power

Wind power has received much attention from viewpoints of the exhaustion problem of fossil fuel and global warming. Wind power generation is not possible at all times because it is strongly dependent on the weather. There are also some places where sufficient wind speed is not available. Therefore, one recent trend is generating hydrogen by using wind energy. If it is possible to transform the wind energy to hydrogen and to preserve the hydrogen in an appropriate way, then it can be kept in a stable state for a long time. In that case, the hydrogen can also be transported easily to any place. This chapter focuses on hydrogen generation from a grid connected wind farm.

6.1 Basic Discussion of Hydrogen

Hydrogen is the most abundant element in the universe, making up 75 % of the mass of all visible matter in stars and galaxies. However, much of the hydrogen is combined with other elements in the form of natural gas (CH_4) and water (H_2O). Small-scale distributed hydrogen production from natural gas is most economical. Advanced natural gas to hydrogen refueling stations are being field evaluated. Statistics say that approximately 90% of the hydrogen produced is obtained from natural resources such as natural gas, coal, and oil through a process called reformation.

On the other hand, electrolyzer technology is available today, but using electricity produced from fossil fuels to make hydrogen creates significant greenhouse gases. However, electrolyzers open the possibility of using electricity made from renewable and nuclear sources to produce carbon-free hydrogen.

Hydrogen has many advantages compared to other energy sources. Hydrogen possesses the highest amount of energy per kilogram of all fuels. It is not toxic. Because hydrogen dissipates quickly in the air, the risk of explosion is low. Moreover, it can be transported safely through a gas pipeline.

However, hydrogen also has some disadvantages when considered as a fuel. Hydrogen has the lowest amount of energy per unit of volume of all fuels. It has a wide range of flammability and it burns in low concentrations. The storage method for hydrogen is not so easy compared to other gas and liquid fuels.

Some applications of hydrogen are listed below.

- Hydrogen energy station: One of the present trends of hydrogen use is in hydrogen energy stations. At present, most hydrogen is produced from natural gases. In the future, renewable energies such as solar and wind energies might be incorporated with the present trend of hydrogen production shown in Fig. 6.1.
- Ammonia production: The production of ammonia uses nitrogen from the air, which is reacted with hydrogen to produce ammonia in the Haber process. Ammonia itself is then primarily used in fertilizer manufacture, but also in industrial refrigeration and the manufacture of a variety of industrial chemicals.
- Metals: Hydrogen is mixed with inert gases to obtain a reducing atmosphere, which is required for many applications in the metallurgical industry, such as heat treating steel and welding. It is often used in annealing stainless steel alloys, magnetic steel alloys, sintering, and copper brazing.
- Chemicals: Hydrogen is used as a raw material in the chemical synthesis of hydrogen peroxide, polymers, and solvents. Hydrogen is used to purify gases (e.g., argon) that contain trace amounts of oxygen, using catalytic combination of the oxygen and hydrogen followed by removal of the resulting water.
- Pharmaceuticals: The pharmaceutical industry uses hydrogen to manufacture vitamins and other pharmaceutical products.
- Oil refining: In oil refineries, hydrogen is used for upgrading the more viscous oil fractions to produce products such as gasoline and diesel and for removing contaminants such as sulfur.
- Glass and ceramics: In float glass manufacturing, hydrogen is required to prevent oxidation of the large tin bath.
- Food and beverages: It is used to hydrogenate unsaturated fatty acids in animal and vegetable oils, producing solid fats for margarine and other food products.
- Electronics: Hydrogen is used as a carrier gas for such active trace elements as arsine and phospine in the manufacture of semiconducting layers in integrated circuits.

- Miscellaneous: Generators in large power plants are often cooled with hydrogen, since the gas has high thermal conductivity and offers low friction resistance.
- Liquid hydrogen is used as a rocket fuel.
- The nuclear fuel industry uses hydrogen as a protective atmosphere in the fabrication of fuel rods.

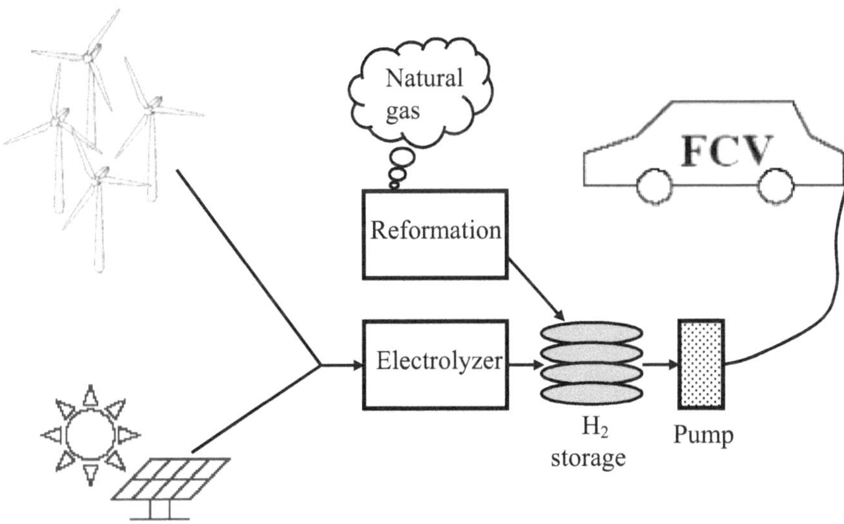

Fig. 6.1 Hydrogen energy station

6.2 Modeling of a Hydrogen Generator

Hydrogen present in nature combined with other elements such as in water or natural gas. Therefore, generation of hydrogen gas means removing it molecules where it is combined with other atoms. Electrolysis is a process that produces hydrogen using water and electric power.

It is well-known that water has two hydrogen and one oxygen atom. When an electric current passes through water, the atoms of hydrogen and oxygen separate from each other. The oxygen goes closer to the positive electrode, which is called the anode. On the other hand, the hydrogen goes toward the negative electrode, which is called the cathode. This can be expressed by simple chemical equation as follows:

$$H_2O = H_2 + \frac{1}{2}O_2 \tag{6.1}$$

Faraday's law for electrolysis states, "The mass of a substance produced at an electrode during electrolysis is proportional to the number of moles of electrons (the quantity of electricity) transferred at that electrode."

In brief, Faraday's law can be expressed by the following equation:

$$\eta = \frac{It}{zF} \tag{6.2}$$

where, I is the current in amperes, t is the time in sec, z is the valence number of ions of the substance (electrons transferred per ion), F is the Faraday constant, and η is the amount of substance ("number of moles") produced.

According to the Faraday's law of electrolysis, it is possible to generate hydrogen gas by using an electrolyzer (ELL). From Eq. 6.2, it is clear that the amont of hydrogen produced depends only on the electrolyzer current. Therefore, the hydrogen generator can be simulated by the electrolyzer and the power electronic devices that will provide constant current to the electrolyzer. To simulate the hydrogen generator, it is necessary to know the electrolyzer characteristics precisely. The characteristics of one electrolyzer are shown in the Appendix.

6.3 Topological Overview

Two types of hydrogen generator topologies which can be used to generate hydrogen by using wind power are presented in the following section.

6.3.1 Hydrogen Generator Model I (Rectifier, DC Chopper, and Electrolyzer)

The schematic diagram of the hydrogen generator model I shown in Fig. 6.2 is composed of a rectifier, a DC chopper, and an electrolyzer. Some portion of the wind farm output is rectified and then a DC chopper supplies constant DC current to the electrolyzer to generate hydrogen gas. This type of hydrogen generator needs a controller to provide constant DC current to the electrolyzer. The hydrogen production is maintained constant at the rated level of the electrolyzer by controlling the DC chopper gate signal, as shown in Fig. 6.3. The error signal between hydrogen generator consumed real power and its reference is progressed through a PI controller and then the chopper duty cycle is generated. The duty cycle of the chopper is compared with the triangular carrier signal and the gate signal for the GTO device of the DC chopper is generated. This type of hydrogen generator is independent of connection point voltage fluctuation because the DC chopper can maintain constant electrolyzer current. Therefore, the fluctuation of the wind farm terminal voltage due to the random variation of the wind may not influence the constant hydrogen generation. But the smoothing capacitor, DC chopper power electronic devices, and their control system definitely increase the system cost.

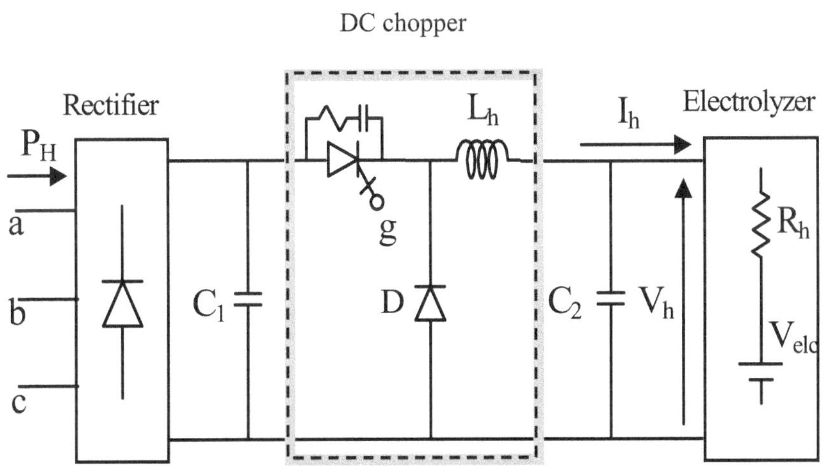

Fig. 6.2 Hydrogen generator model system I

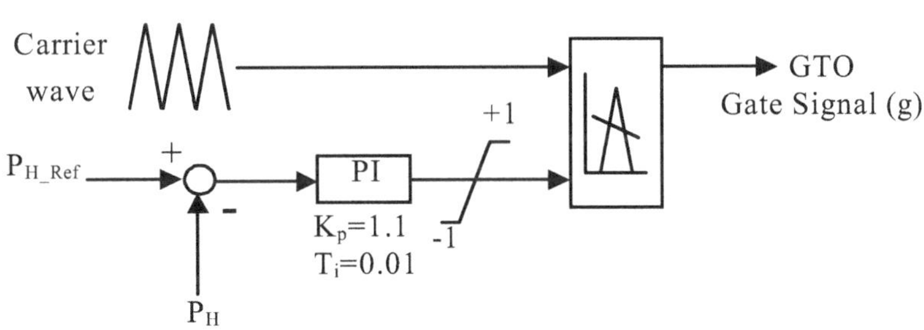

Fig. 6.3 Control block of the DC chopper

6.3.2 Hydrogen Generator Model II (Rectifier and Electrolyzer)

The schematic diagram of the hydrogen generator model II shown in Fig. 6.4 is composed of a rectifier and an electrolyzer. Some portion of the wind farm output is rectified and then the DC current enters the electrolyzer directly to generate hydrogen gas. The voltage fluctuation at the hydrogen generator connection point can influence the volume of hydrogen generated. Therefore, constant hydrogen cannot be generated from this hydrogen generator if the voltage at the connection point cannot be maintained constant. The installation cost of the hydrogen generator model II is comparatively lower than that of the hydrogen generator model I. Additionally, efficiency would be somewhat better than that of the hydrogen generator model I because one power conversion step (DC-to-DC) can be eliminated. However, this type of topology may be suitable for wind power application, only if the hydrogen generator connection point voltage can be kept constant. Otherwise, the fluctuating electrolyzer current may damage the electrolyzer unit.

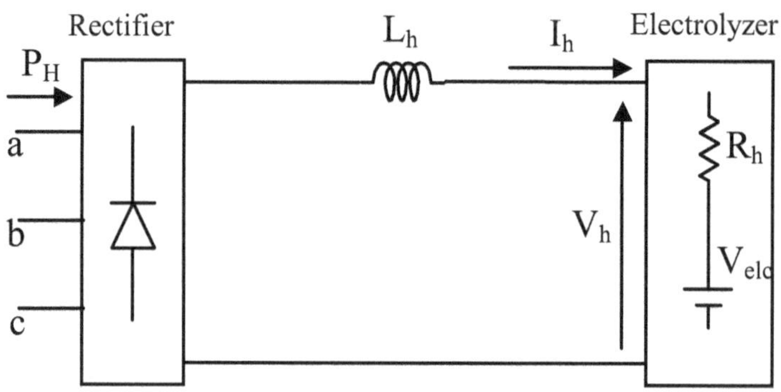

Fig. 6.4 Hydrogen generator model system II

6.3.3 A Method for Calculating the Amount of Hydrogen Gas

When simulating the hydrogen generator, then it is also necessary to know how much hydrogen is generated from the hydrogen generator. Here, one method is described in brief.

From the one cell electrolyzer parameter, first the hydrogen gas generated per second, Q, is calculated as follows:

$$Q = \frac{7.5 I_h}{3600 I_{hn}} \quad (\mathrm{Nm^3/s}) \tag{6.3}$$

where, I_{hn} is the nominal value of one electrolyzer cell.

Then, from Eq. 6.3, the total hydrogen gas generated in T sec can be calculated from Eq. 6.4:

$$Q_{Total} = \frac{7.5}{3600 I_{hn}} \int_0^T I_h(t)dt \quad (Nm^3) \tag{6.4}$$

6.4 Recent Trend in Hydrogen Generation from Wind Power

Hydrogen can be generated from both a fixed and variable speed wind turbine generator systems (WTGS). Both stand-alone and grid connected hydrogen generation systems can be adopted with a WTGS. In Figs. 6.5 and 6.6, the schematic diagrams are presented for hydrogen generation from a grid connected wind farm and a stand-alone WTGS, respectively.

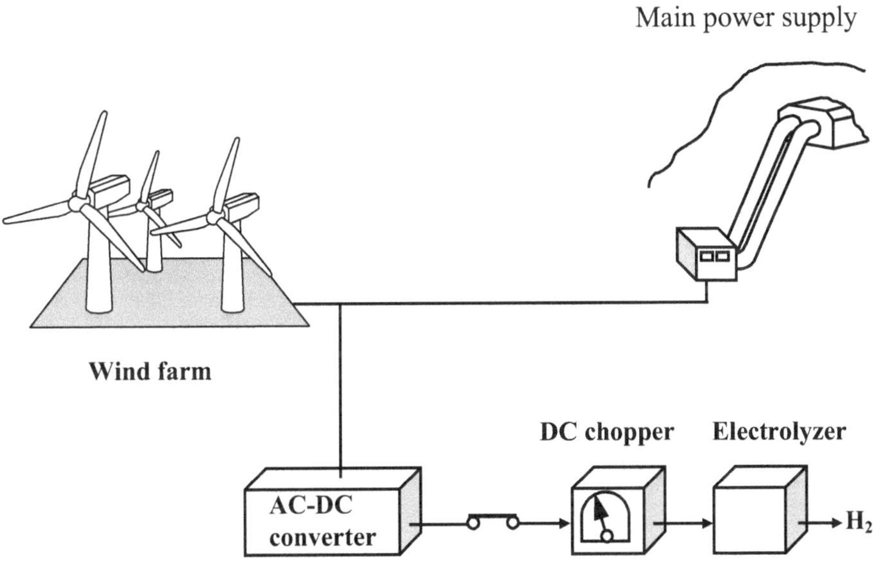

Fig. 6.5 Schematic diagram of hydrogen production from agrid connected wind farm

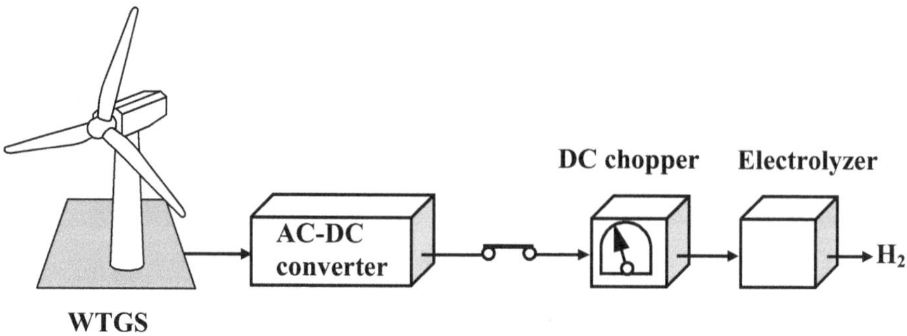

Fig. 6.6 Schematic diagram of hydrogen production from a stand-alone WTGS

6.5 Hydrogen Storage in a Wind Turbine Tower

Hydrogen can be stored in different ways. One of the ways to store the hydrogen gas generated using wind turbine power is inside the tower itself. The tower of the wind turbine, in general, possesses a rigid structure. Therefore, it might be somewhat cost-effective if hydrogen gas generated using wind power can be stored inside the tower of wind turbine. National Renewable Energy Laboratory (NREL) has done an excellent study on this topic.[1] This section is presented briefly in the light of the NREL report on hydrogen storage in wind turbine towers [138].

6.5.1 Conventional Pressure Vessels

Industrial pressure vessels are often built of carbon steel similar to that used in turbine tower construction. Although the most economical pressure vessel geometry is long and slender, vessels are often limited by shipping constraints to a practical length of about 25 meters. This length limitation means that in order to better distribute the high fixed costs associated with nozzles and manways, pressure vessels are designed with relatively large diameters and high pressure ratings. Although higher pressures reduce the cost per kilogram of stored gas, higher pressures require additional compression costs.

[1] To learn about National Renewable Energy Laboratory (NREL), visit at http://www.nrel.gov/

6.5.2 Conventional Towers

The 1.5-MW tower model specified in the WindPACT Advanced Wind Turbine Designs Study is chosen as the baseline conventional tower.[2] The modeled tower is shown in Fig. 6.7.

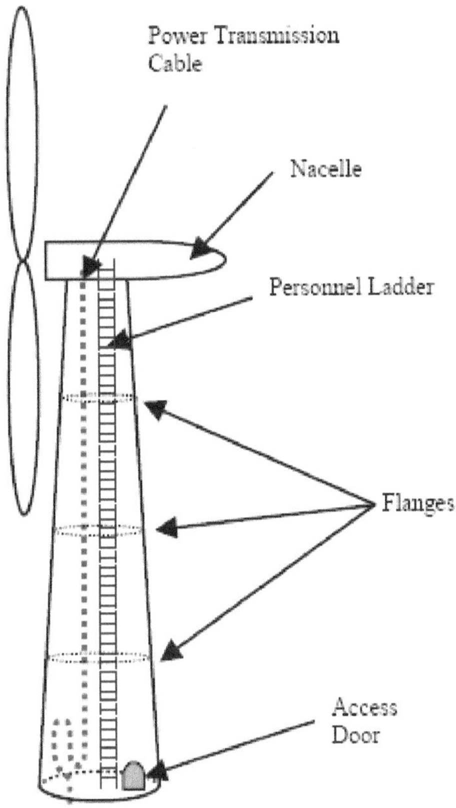

Fig. 6.7 Baseline tower model (This figure was developed by the National Renewable Energy Laboratory for the U.S. Department of Energy [138])

---

[2] The Wind Partnerships for Advanced Component Technology (WindPACT) was started in 1999 to assist industry in lowering the cost of energy by designing and testing innovative components, such as advanced blades and drive trains. For more detail, visit the NREL homepage or directly at http://www.nrel.gov/wind/advanced_technology.html

6.5.3 Hydrogen Tower Considerations

Hydrogen storage creates a number of additional considerations in turbine tower design. Under certain conditions, hydrogen tends to react with steel, adversely affecting several of steel's engineering properties, including ductility, yield strength, and fatigue life. Additionally, storing hydrogen at pressure significantly increases the stresses on the tower; therefore, storing hydrogen under pressure is likely to require wall reinforcement. These factors require a structural analysis to evaluate how internal pressure may affect the tower's design life.

However, details of the structural analysis reported in [138] are not presented in this section. Only the conceptual design of hydrogen storage inside the tower itself is presented as follows.

The most straightforward design concept is illustrated in Fig. 6.8.

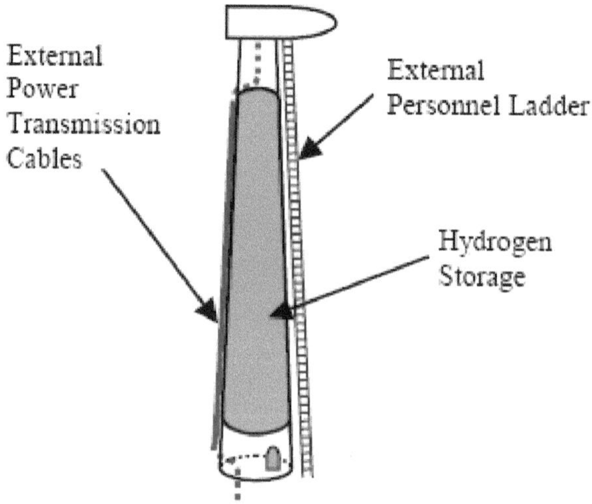

Fig. 6.8 Full hydrogen tower (This figure was developed by the National Renewable Energy Laboratory for the U.S. Department of Energy [138])

This design places pressure head weldments near the top and bottom of the tower and moves the access ladder and power transmission lines to the exterior of the tower. If the power transmission lines are moved to the outside, then they must be protected by conduit. This prevents standard droop-cable design and requires leaving 9 m of space in the tower above the upper pressure head to allow for installation of cable with torsional flexibility.[3] The bottom end cap allows the equipment that is normally stored in the base of a tower to remain there. These pressure heads also contain the pressure vessel loads, which allow the foundation and nacelle design to be unaffected by hydrogen storage. This concept is appealing because it offers a great amount of hydrogen storage with relatively simple design modifications. For these reasons, this idea stands out as a cost-effective option.

Two other designs considered are variations of this first concept. One of them requires a small pipe running down the axis of the turbine, which would allow the use of standard power transmission cables (Fig. 6.9).

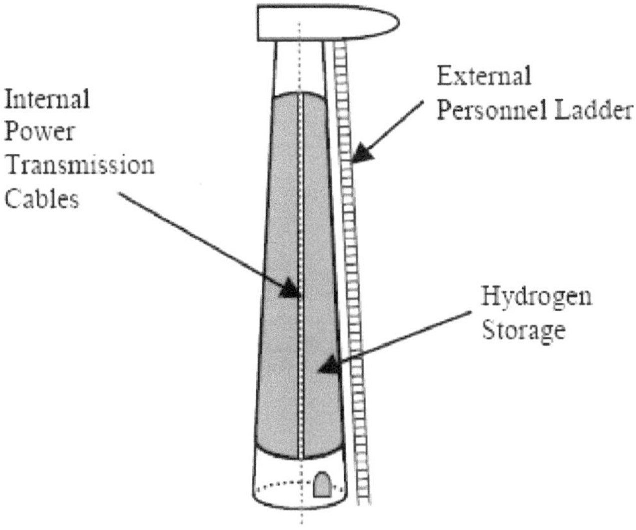

Fig. 6.9 Hydrogen tower with internal power cable (This figure was developed by the National Renewable Energy Laboratory for the U.S. Department of Energy [138])

This idea was driven by an early interest in modifying the tower design as little as possible. Although this pipe installation may add cost and complexity to construction methods, the cost would be partially offset because it would allow a standard power cable design, eliminating the need for the high-flex cable and the conduit required for the exterior power transmission design.

---

[3] Poore, R.; Lettenmaier, T. (2002). Alternative Design Study Report: WindPACT Advanced Wind Turbine Drive Train Designs Study. NREL/SR-500-33196. Work performed by Global Energy Concepts, LLC, Kirkland, Washington. Golden, CO: National Renewable Energy Laboratory.

Another variation of the full hydrogen tower concept requires a pipe large enough to accommodate the personnel ladder. It is possible that in some areas an external ladder will cause problems with bird perching or ice formation. This concept was therefore attractive because it eliminates the need for an external ladder and conduit. However, this concept sacrifices about 10% of the hydrogen storage volume, further raising the cost/mass ratio. The added cost and construction complexity of the large diameter pipe makes this concept less attractive than other options. It was therefore excluded from further consideration.

The significant cost associated with the bottom end cap motivated yet another design concept. Instead of manufacturing the large bottom end cap, a thin plate could be welded flush with the bottom of the tower. This plate acts as a seal but isn't designed to bear a load. The pressure load would be borne by the foundation and the flange bolts at the base of the tower (Fig. 6.10).

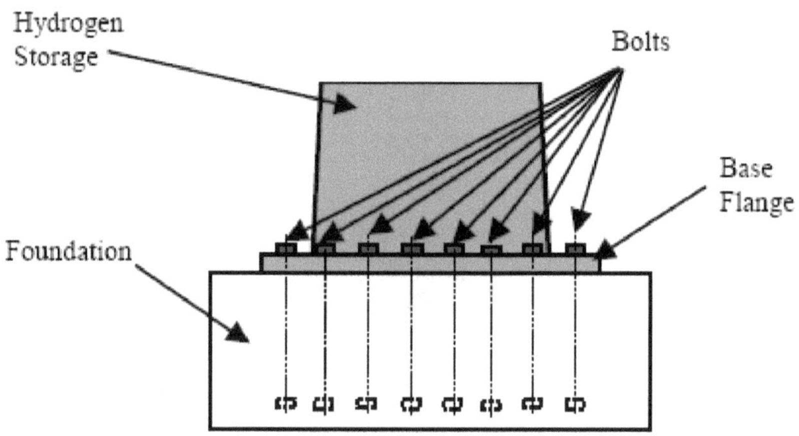

Fig. 6.10 Hydrogen tower, alternate foundation design (This figure was developed by the National Renewable Energy Laboratory for the U.S. Department of Energy [138])

This concept offers marginally greater hydrogen storage capacity but creates a large bending moment on the foundation. The pressurized hydrogen pushes down in the middle of the foundation, and the pre-stressed bolts pull up around the foundation's perimeter. Also, the power electronics and wind turbine control equipment normally stored in the base of the tower would have to be stored elsewhere (either in a building adjacent to the tower's base or in the nacelle). If, in the future, hydrogen towers become a part of the energy economy, this foundation concept would be a good subject for further study.

The final major concept uses only a section of the tower for hydrogen storage. This option is appealing for several reasons:

It allows for standard electrical cable and personnel access installation above the storage space; it keeps the upper exterior of the tower clear of the ladder and conduit, which would otherwise require additional blade clearance to prevent strikes; and finally, it allows the option of scaling down the total cost and storage capacity, which, depending on the application, may be desirable. This concept does, however, result in a higher cost/mass ratio because it brings the end caps closer together, moving away from the ideal long, slender shape.

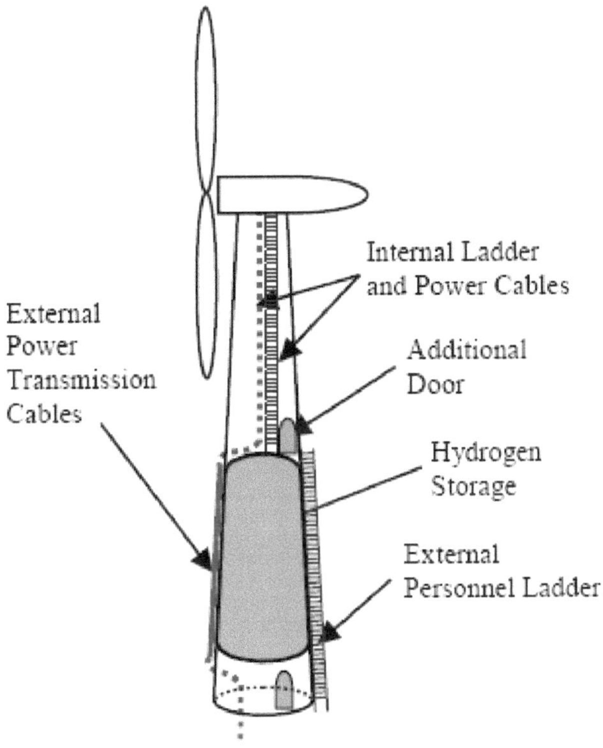

Fig. 6.11 Storage in the base of a hydrogen tower (This figure was developed by the National Renewable Energy Laboratory for the U.S. Department of Energy [138])

6.6 Chapter Summary

Hydrogen is going to be one of the vital sources of energy in the near future. Different types of uses of hydrogen are presented in this chapter. Due to the growing interest in hydrogen, it is becoming the concern of many companies and laboratories to produce hydrogen in the most economical way. One of the recent trends in hydrogen production is to use wind power. Hydrogen generation system can be adopted at both a stand-alone and a grid connected WTGS. In this chapter, detailed modeling and control strategy for a hydrogen generator are presented, including the electrolyzer modeling. At the end of this chapter, some interesting concepts of hydrogen storage inside the wind turbine are presented in the light of the NREL report.

Chapter 7

Wind Farm Operating Strategy with an Energy Capacitor System and a Hydrogen Generator

In this chapter, the design and control strategy for a wind farm composed of wind generators, a hydrogen generator (HG), and an energy capacitor system (ECS) are presented. One of the recent challenges to wind power generation is smoothing the fluctuation in wind generator output power due to the random variation of the wind speed. This chapter proposes an energy capacitor system (ECS), composed of power electronic devices and an electric double layer capacitor (EDLC), to smooth the line power of a wind farm of fixed speed wind generators.

A constant output power reference is not a good choice because sometimes the wind speed is very low and then sufficient power cannot be obtained. In that case, an energy storage device can solve the problem, but large energy capacity may be needed. Therefore, an exponential moving average (EMA) is proposed to generate the reference output power, and thus the energy capacity of the ECS unit can be small.

Another salient feature of this chapter is the generation of hydrogen by using wind energy. Hydrogen has received much attention in recent years as a new energy source. Two types of hydrogen generators are considered and their merits and demerits are analyzed. By taking advantage of an ECS, the cost and performance effective topology of a hydrogen generator is proposed. Detailed control strategies for a hydrogen generator and an energy capacitor system are discussed.

In addition, the transient stability augmentation of a wind farm by using an ECS is analyzed. It is reported that an ECS can enhance the low voltage ride through (LVRT) capabilities of each wind generator of a wind farm. Moreover, it can enhance the transient stability of the power system. The effectiveness of the proposed system is verified by a simulation analysis using PSCAD/EMTDC [126]. The schematic diagram of the cooperative control system among a wind farm, an ECS, and a hydrogen generator is shown in Fig. 7.1.

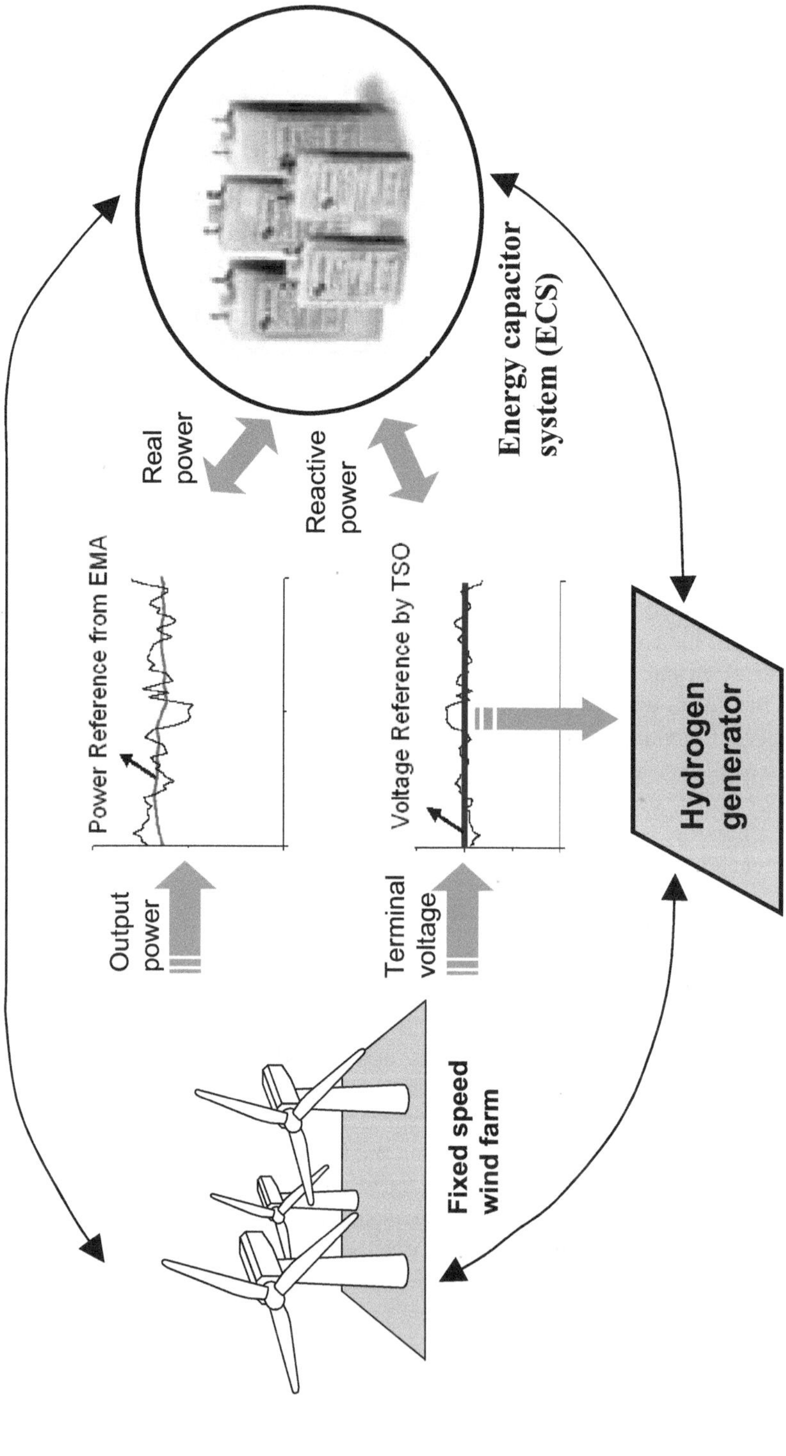

Fig. 7.1 Cooperative control among a wind farm, an ECS, and a Hydrogen generator

7.1 Modeling and Control Strategy for an Energy Capacitor System

The energy capacitor system (ECS) consists of an EDLC and power electronic devices used as an energy storage system (ESS). The schematic diagram of an ECS is shown in Fig. 7.2, where the EDLC bank is shown by the rectangular box, the PWM voltage source converter (VSC) is shown by the dotted line, and the DC DC buck/boost converter is shown by the dashed line. The PWM VSC controls the DC-link voltage and the reactive power flowing from the grid, whereas the DC-DC buck/boost converter controls the real power. The individual component modeling of an ECS is presented in this section [114].

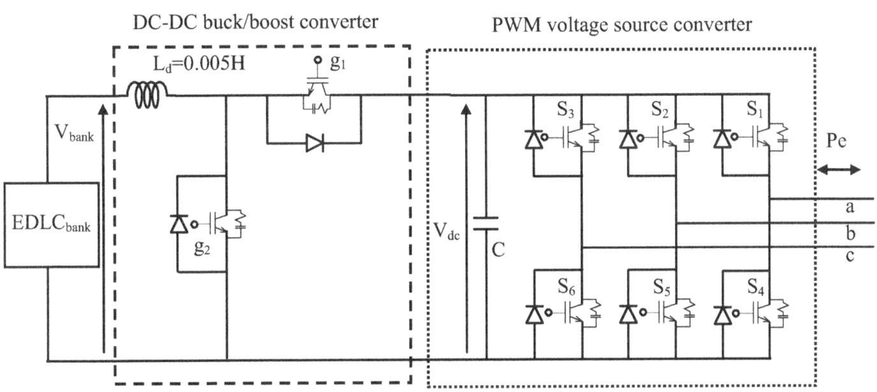

Fig. 7.2 Schematic diagram of anenergy capacitor system (ECS) [114]

7.1.1 EDLC Modeling

In this analysis, a distributed model of an EDLC cell is considered because it can express the terminal characteristic precisely. The parameters of a single EDLC cell for the lumped and distributed models are shown in Tables 7.1 and 7.2, respectively. The rated EDLC bank voltage chosen is 5.0 kV. At the end of 2007, an EDLC unit rated at 6.6 kV is available in the power industry. In the simulations, it is assumed that 1850 EDLC cells are connected in series to make a string with a 5.0 kV voltage rating. The balancing circuits are neglected here for simplicity, though it is necessary to connect many EDLC cells in series in practical applications. The rated capacity of the ECS is 20 MW, 0.305 MWh. To obtain such energy, 54 strings are needed to work in parallel. After the circuit simplification, the combined distributed parameters of the EDLC bank can be obtained, as presented in Table 7.3.

Table 7.1 Lumped model parameters of an EDLC cell

Rated Voltage	2.7 V
Capacitance, Cb	3000 F
Internal Resistance, Rb	9 mΩ

Table 7.2 Distributed model parameters of an EDLC cell

Capacitance		Internal Resistance	
Cb1	60 F	Rb1	0.36 mΩ
Cb2	1500 F	Rb2	9.0 mΩ
Cb3	1440 F	Rb3	8.64 mΩ

Table 7.3 Distributed model parameters of an EDLC bank

Capacitance		Internal Resistance	
C_{b1}	1.76F	R_{b1}	0.012Ω
C_{b2}	44.00F	R_{b2}	0.308Ω
C_{b3}	42.24F	R_{b3}	0.295Ω

7.1.2 Modeling and Control Strategy of a VSC

In this analysis, the well-known cascaded control scheme with independent control of the active and reactive current is developed, as shown in Fig. 7.3. The aim of the control is to maintain the magnitude of voltage at the wind farm terminal at the desired reference level under randomly fluctuating wind speed conditions. The DC-link voltage is also kept constant at the rated value. Finally, the three-phase reference signals are compared with the triangular carrier wave signal to generate the switching signals for the IGBT switched VSC. A GTO gate device can also be adopted instead of the IGBT. High switching frequencies can be used to improve the efficiency of the converter without incurring significant switching losses.

In the simulation, the switching frequency chosen is 1000 Hz. The snubber circuit resistance and capacitance values of the IGBT devices are 5000 Ω and 0.05 μF, respectively. The DC-link voltage is 5.0 kV. The ECS is connected to the 66 kV line through a single step down transformer (66 kV/2.72 kV) with 0.2 p.u leakage reactance (base value 100 MVA). The DC-link capacitor value is 20000 μF. The detailed modeling and control strategy for a PWM based VSC are available in Chap. 4.

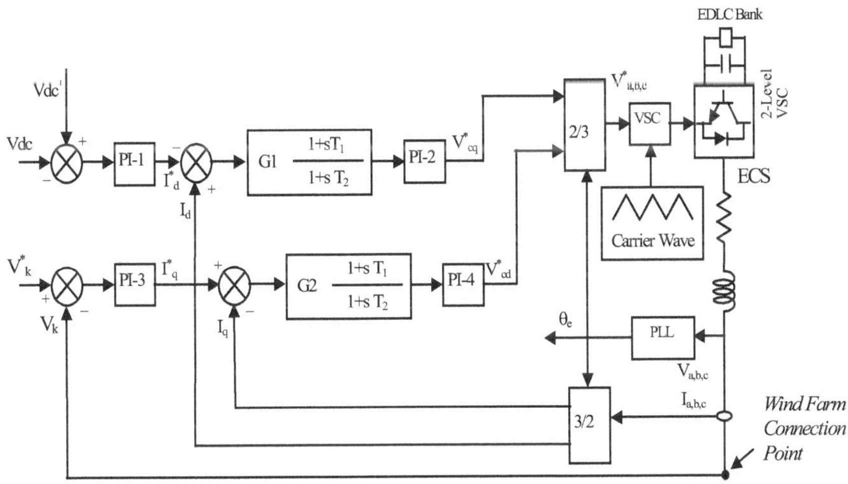

Fig. 7.3 Control block diagram of PWM based VSC [114]

7.1.3 Modeling of a DC-DC Buck/Boost Converter

The DC-DC buck/boost converter shown inside the dashed line of Fig. 7.4 operates by alternately controlling switches g_1 and g_2 to be ON or OFF. When the wind farm line power, P_L, is less than the reference power, the EDLC discharges, working in boost converter mode and vice versa. The error signal between the line power and reference power is progressed through a PI controller and then compa-

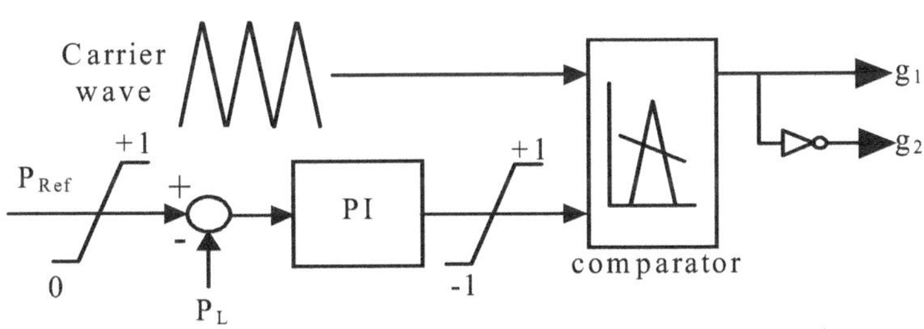

Fig. 7.4 Control block for a DC-DC buck/boost converter [114]

red with the triangular carrier wave to generate the gate signals for the buck/boost converter, as shown in Fig. 7.2. When a network disturbance occurs in the the power system, then the DC-DC buck/boost converter might be forced to work in the charging mode only. Therefore, it can store the transient energy of the power system and can enhance the transient stability of the rest of the system. The frequency of the triangular carrier signal chosen is 250 Hz.

7.2 Hydrogen Generator Model System

Recently, hydrogen is considered one of the alternative energy sources. According to the Faraday's law of electrolysis, it is possible to generate hydrogen gas by using an electrolyzer (ELL). In this analysis, the electrolyzer is used for hydrogen production, as explained in Chap. 6. The electrolyzer characteristics used in this analysis are shown in the Appendix. Two types of hydrogen generator topologies are used in this study for constant hydrogen generation, as described below.

In this analysis, the hydrogen generator composed of a rectifier, a DC chopper, and an electrolyzer is called HG-I. Hydrogen production is maintained constant at 10 MW by controlling the DC chopper gate signal, as shown in the control block of the DC chopper used in Chap. 6. The error signal between hydrogen generator consumed real power and its reference is progressed through a PI controller, and then the chopper duty cycle is generated. The duty cycle of the chopper is compared with the triangular carrier signal, and the gate signal for the GTO device of the DC chopper is generated. The triangular carrier frequency chosen is 450 Hz. In this analysis, the lumped model of an electrolyzer is used for the simulation. The capacity of the individual electrolyzer cell is assumed to be 44.1 kW. One string consists of 10 cells. The lumped model consists of 23 strings working in parallel to ensure sufficient electrolytic current. The parameters of the individual cell and lumped model of the electrolyzer are shown in Tables 7.4 and 7.5, respectively.

Table 7.4 Specifications of one electrolyzer cell

Rated power consumption	44.1 (kW)
Rated voltage	107.5 (V)
Hydrogen gas volume	7.5 (Nm^3)/hr
Resistance	0.031 (Ω)
DC source	94.8 (V)

Table 7.5 Lumped model parameters of an electrolyzer

Inductance L_h	3 (mH)
Filtering capacitor C_1	20000 (μ F)
Filtering capacitor C_2	1000 (μ F)
Resistance R_h	0.0135 (Ω)
DC source V_{elc}	948 (V)

The hydrogen generator composed of a rectifier and an electrolyzer is called HG-II. This model is the simplest one and the details of this model are available in Chap. 6 of this book. In this case, the same lumped model parameters of the electrolyzer used in HG-I are used, as shown in Table 7.5.

7.3 Wind Farm Output Power Smoothing and Terminal Voltage Regulation

7.3.1 Model System

Figure 7.5 shows the model system used for the simulation analyses of the fixed speed wind farm output power smoothing and terminal voltage regulation. Here, one synchronous generator (SG) is connected to an infinite bus through a transformer and a double circuit transmission line. One wind farm (50 MVA) composed of fixed speed wind generators is connected to a network via a transformer and short transmission line. In this analysis, for the sake of precise analysis, a real wind park model is considered instead of an aggregated wind park model. A capacitor bank has been used for reactive power compensation at steady state, as described in Chap. 2. The wind turbine characteristic used in this analysis is also described in Chap. 2. The conventional pitch controller is used with a wind turbine, as described in Sect. 3.1 of Chap. 3. The ECS and hydrogen generator are connected to point K, as shown in Fig. 7.5. Both hydrogen generator models (HG-I and HG-II) are used in the simulation. The AVR (automatic voltage regulator) and GOV (governor) control system models for the synchronous generator described in Sect. 2.3.4.1 of Chap. 2 are used in this analysis. The generator parameters shown in Table 7.6 are used. The system base is 100 MVA.

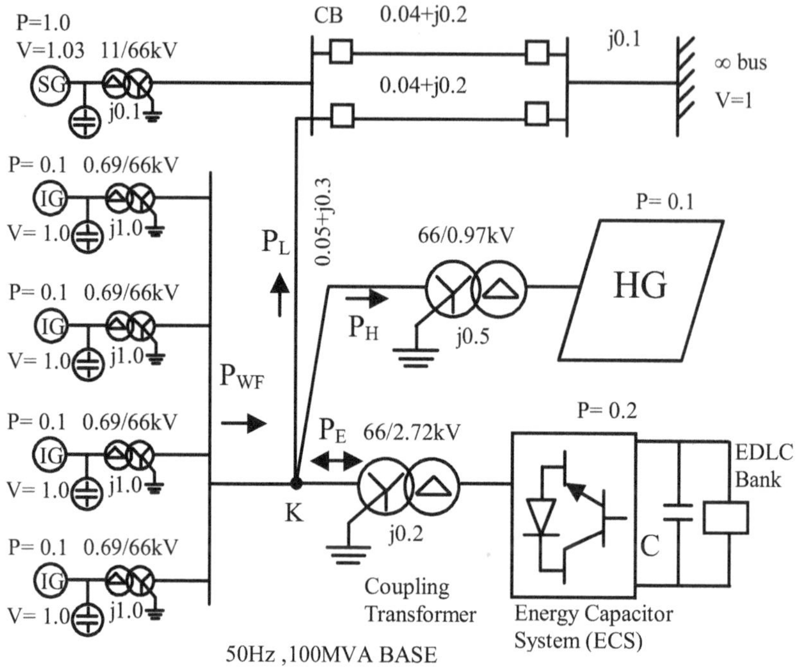

Fig. 7.5 Model system

Table 7.6 Generator parameters

SG		IG	
MVA	100	MVA	10
Ra (pu)	0.003	r1 (pu)	0.01
Xa (pu)	0.13	x1 (pu)	0.1
Xd (pu)	1.2	Xmu (pu)	3.5
Xq (pu)	0.7	r21 (pu)	0.035
Xd′ (pu)	0.3	x21 (pu)	0.030
Xq′ (pu)	0.22	r22 (pu)	0.014
Xd″ (pu)	0.22	x22 (pu)	0.098
Xq″ (pu)	0.25	H (sec)	1.5
Tdo′ (sec)	5.0		
Tdo″ (sec)	0.04		
Tqo″ (sec)	0.05		
H (sec)	2.5		

7.3.2 Determination of Output Line Power Reference, P_{Ref}

One objective of this analysis is to smooth the wind farm line power, P_L, as shown in Fig. 7.5. The reference, P_{Ref}, is generated from the exponential moving average (EMA) of the power difference between the wind farm output, P_{WF}, and the consumed power of the hydrogen generator, P_H. The formula for an exponential moving average is shown in Chap. 3. Here, a 180 sec (60 periods each of 3 sec) EMA is used to generate the line power reference, P_{Ref}. For the first period EMA calculation, the average value is used. Therefore, the simulation results for the first 180 sec are not shown. The ECS will supply/absorb the necessary/surplus real power according to the error signal between P_{Ref} and P_L by using a DC-DC buck/boost converter, as shown in Fig. 7.4.

7.3.3 Simulation Study with a WTGS, an ECS, and a Hydrogen Generator

The real wind speed data shown in Fig. 7.6, which were obtained on Hokkaido Island, Japan, are used for each wind generator of the wind farm. The time step and simulation time chosen were 0.00005 sec and 600 sec, respectively. The simulation was done by using PSCAD/EMTDC [126]. The parameters of the PI controllers used in the VSC of Fig. 7.3 are shown in Table 7.7. The proportional gain and integral time constant of the PI controller used in the DC-DC buck/boost converter shown in Fig. 7.4 are 1.0 and 0.05 respectively. Three cases are considered to show the effectiveness of integrating an ECS with wind a farm for line power smoothing and constant hydrogen generation from wind energy.

Table 7.7 The parameters of the PI controllers used in Sect. 7.3

	PI-1	PI-2	PI-3	PI-4
K_p	4.0	0.04	4.0	0.01
T_i	0.1	0.5	0.1	0.5

Case 1: In this case, the performance of HG-I that consists of a rectifier, a DC chopper, and an electrolyzer is demonstrated. The ECS is not considered in this case. The line power and terminal voltage of the wind farm shown in Figs. 7.7 and 7.8, respectively are fluctuating due to wind speed fluctuations. But the DC chopper provides constant DC current to the electrolyzer according to its control strategy explained in Chap. 6. The current of the electrolyzer is shown in Fig. 7.9. Therefore, constant hydrogen generation is possible, though the wind farm termi-

nal voltage is fluctuating. The real power consumption by the hydrogen generator and the total generated hydrogen gas are shown in Figs. 7.10 and 7.11, respectively. The drawbacks of this hydrogen generator topology are the higher installation cost and somewhat lower efficiency due to the loss in DC-DC power conversion.

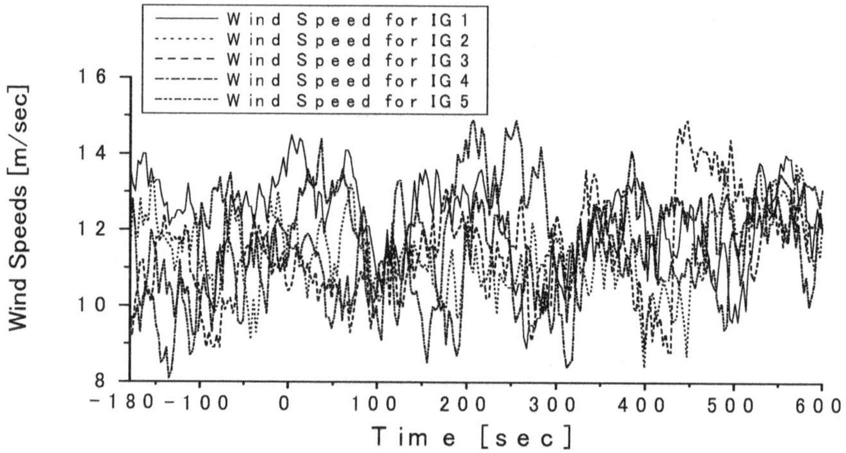

Fig. 7.6 Wind speeds for IG1-IG5

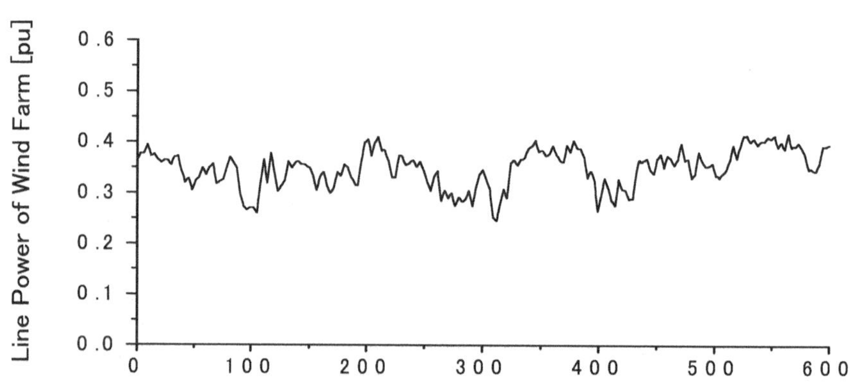

Fig. 7.7 Line power of the wind farm (Case 1)

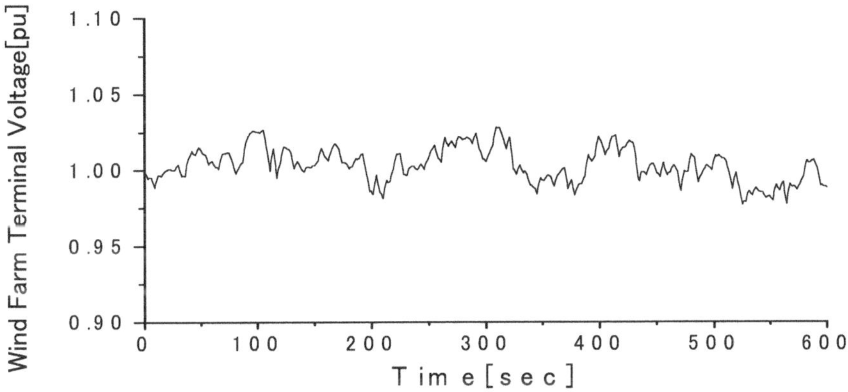

Fig. 7.8 Terminal voltage of the wind farm (Case 1)

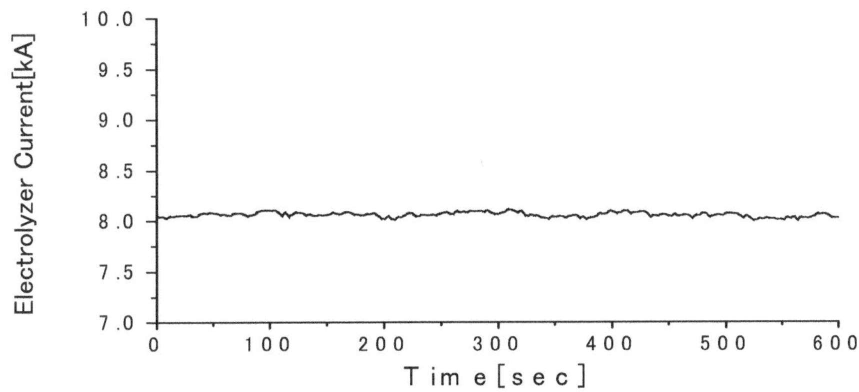

Fig. 7.9 The current of the electrolyzer (Case 1)

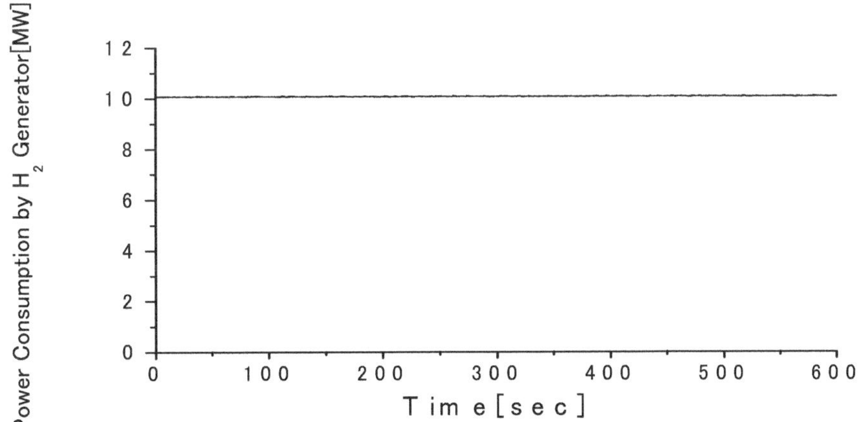

Fig. 7.10 Real power consumption by the H_2 Generator (Case 1)

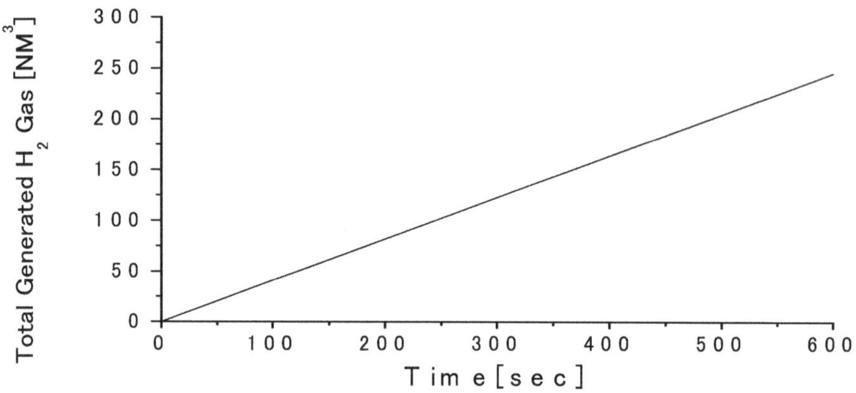

Fig. 7.11 Total generation of H_2 gas (Case 1)

Case 2: In this case, the HG-II that consists of a rectifier and an electrolyzer is considered connected at point K of Fig. 7.5. The ECS is also not considered in this case. Because the wind speed is always fluctuating, the line power and terminal voltage of the wind farm at point K of Fig. 7.5 are fluctuating, as shown in Figs. 7.12 and 7.13, respectively. Therefore, the DC current flowing to the electrolyzer is also fluctuating, as shown in Fig. 7.14. The real power consumption by the hydrogen generator and the total generated hydrogen gas are shown in Figs. 7.15 and 7.16, respectively. It is seen from the simulation results that constant hydrogen production is not possible by using this hydrogen generator topology. From Fig. 7.15, it is clear that the HG-II is consuming more than its rated power, which may damage the electrolyzer.

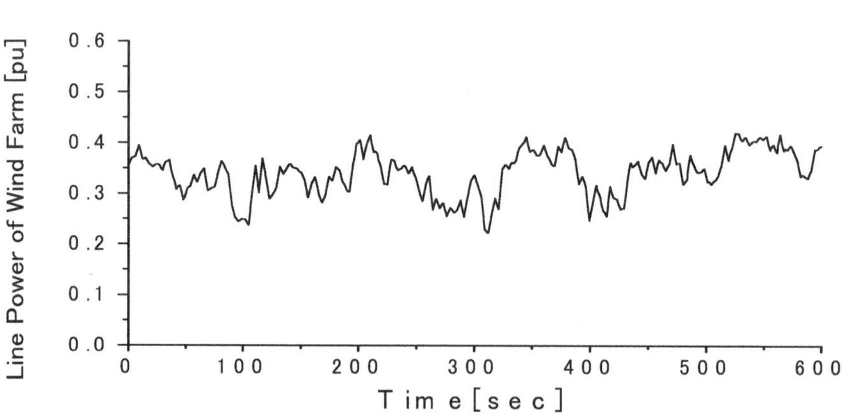

Fig. 7.12 Line power of the wind farm (Case 2)

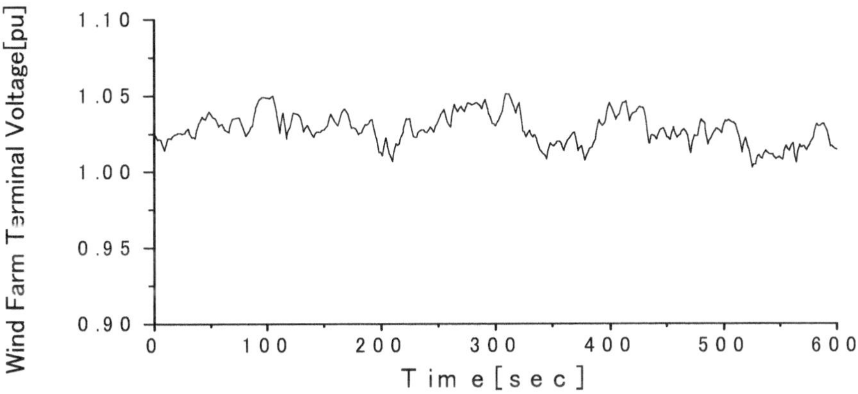

Fig. 7.13 Terminal voltage of the wind farm (Case 2)

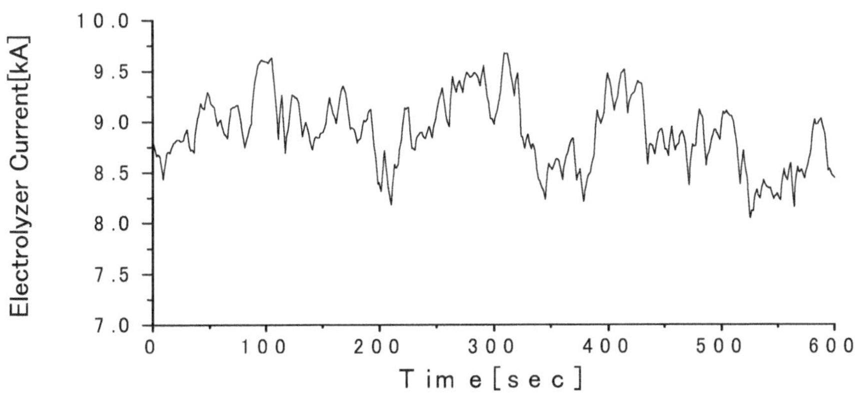

Fig. 7.14 The current of the electrolyzer (Case 2)

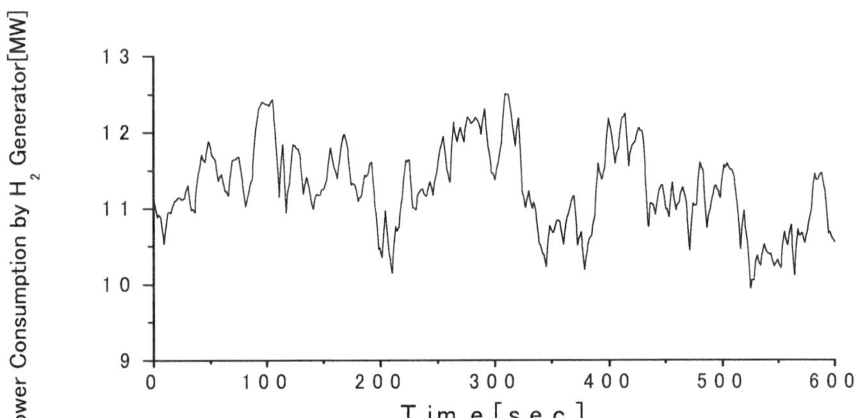

Fig. 7.15 Real power consumption by the H_2 Generator (Case 2)

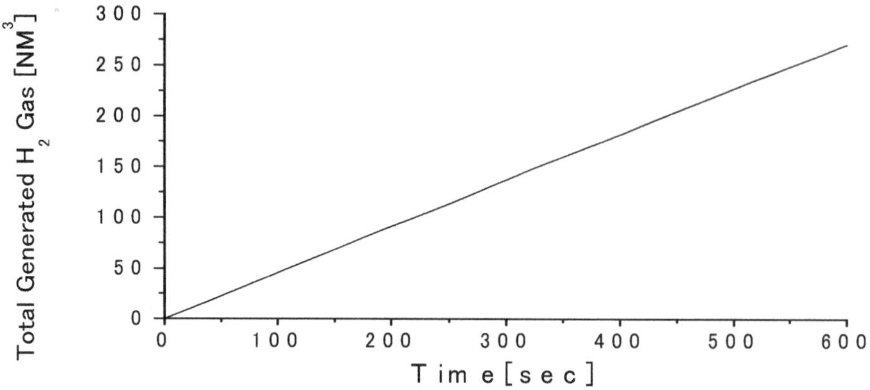

Fig. 7.16 Total generation of H_2 gas (Case 2)

Case 3: In this case, both the ECS and the HG-II composed of a rectifier and an electrolyzer are considered connected at point K of Fig. 7.5. The ECS can regulate the terminal voltage of the wind farm, as shown in Fig. 7.17, by providing or absorbing reactive power at the connection point. The reactive power of the ECS is shown in Fig. 7.18. By using the advantage of constant wind farm terminal voltage, the most economical hydrogen generator topology (HG-II) can be adopted. The electrolyzer current is shown in Fig. 7.19. The real power consumption and total hydrogen generated by this system are shown in Figs. 7.20 and 7.21, respectively. It is seen clearly that at this time the hydrogen generator with a rectifier and an electrolyzer can generate almost constant hydrogen when an ECS is used. But it is not possible when an ECS is not used, as mentioned in Case 2. Due to the fluct-

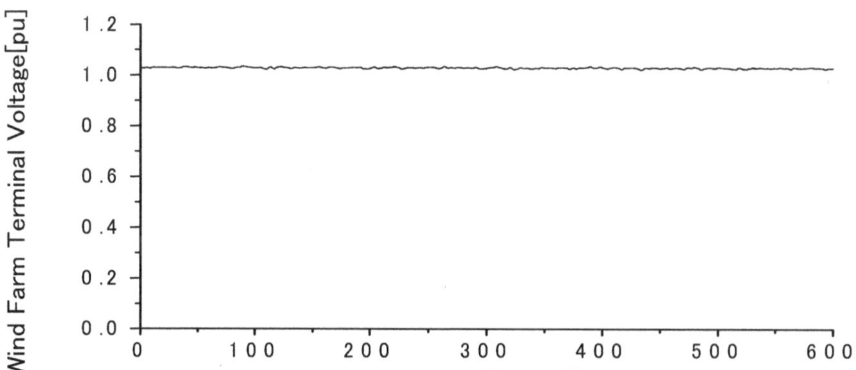

Fig. 7.17 Terminal voltage of wind farm (Case 3)

uation of wind speed,the total generated power of the wind farm is also fluctuating, as shown in Fig. 7.22. The wind farm line power reference calculated using the exponential moving average (EMA) is also shown in the same figure. The ECS will provide the necessary real power to follow the line power reference. As a result, the smoothed line power can be obtained, as shown in Fig. 7.23. The real power of the ECS is also shown in that figure. Therefore, the objective of the proposed system with smoothed line power and constant hydrogen generation can be achieved by using the cost-effective topology. The DC-link voltage, EDLC bank voltage, and stored energy of the EDLC bank are shown in Figs. 7.24 – 7.26, respectively.

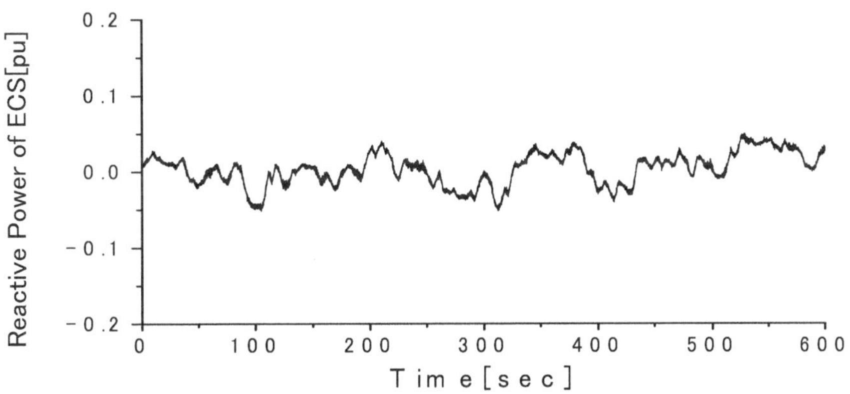

Fig. 7.18 Reactive power of the ECS (Case 3)

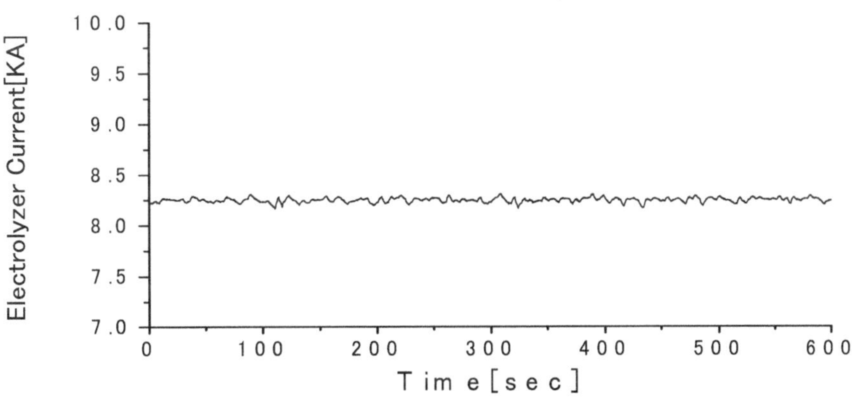

Fig. 7.19 The current of the electrolyzer (Case 3)

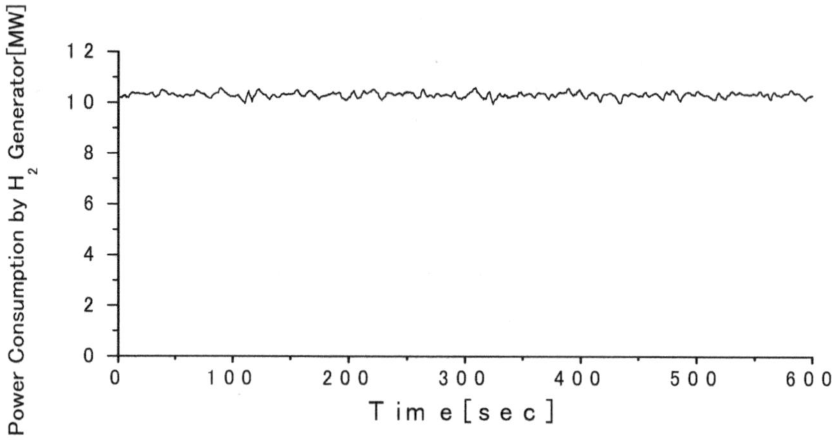

Fig. 7.20 Real power consumption by the H_2 Generator (Case 3)

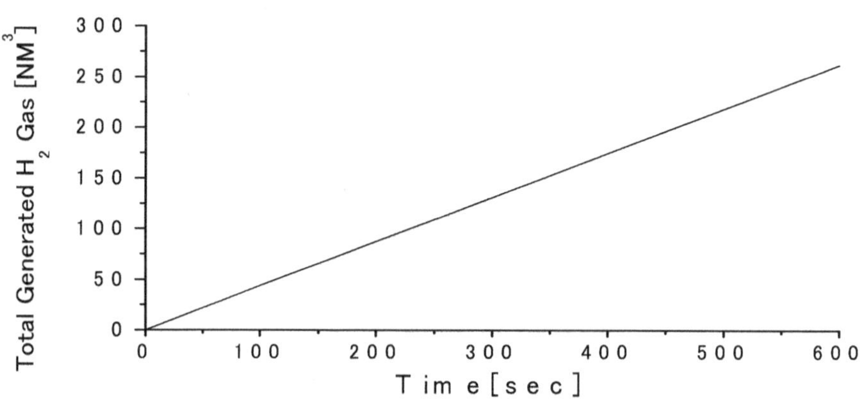

Fig. 7.21 Total generation of H_2 gas (Case 3)

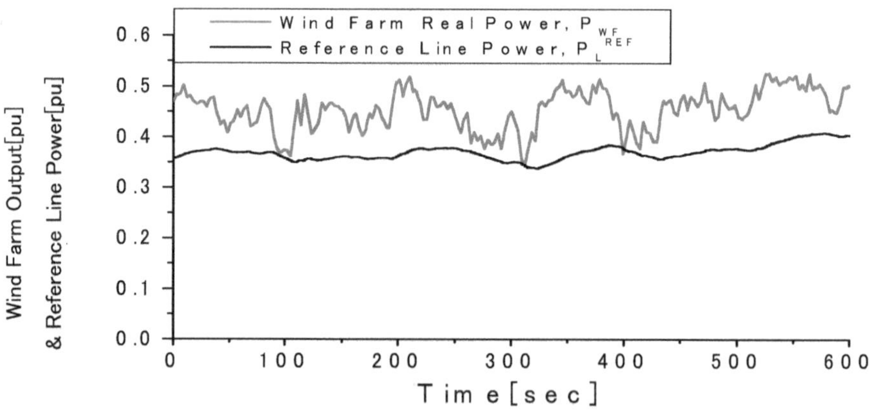

Fig. 7.22 Wind farm real power and reference line power (Case 3)

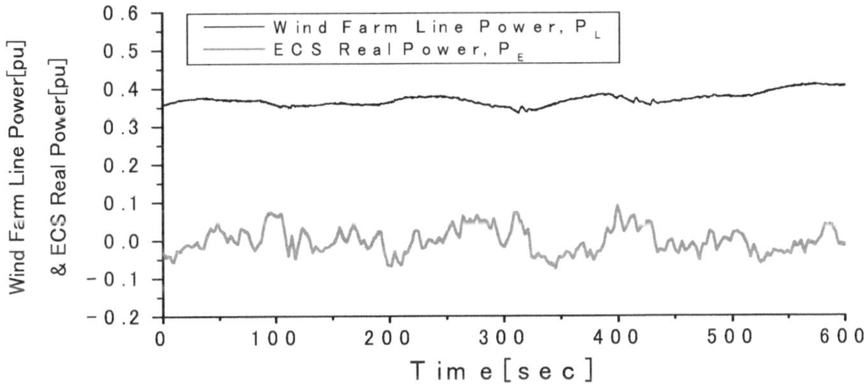

Fig. 7.23 Wind farm line power and ECS real power (Case 3)

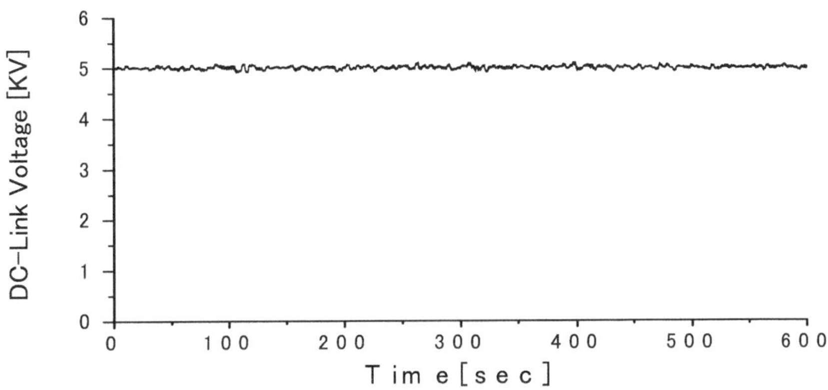

Fig. 7.24 DC-link voltage of the VSC (Case 3)

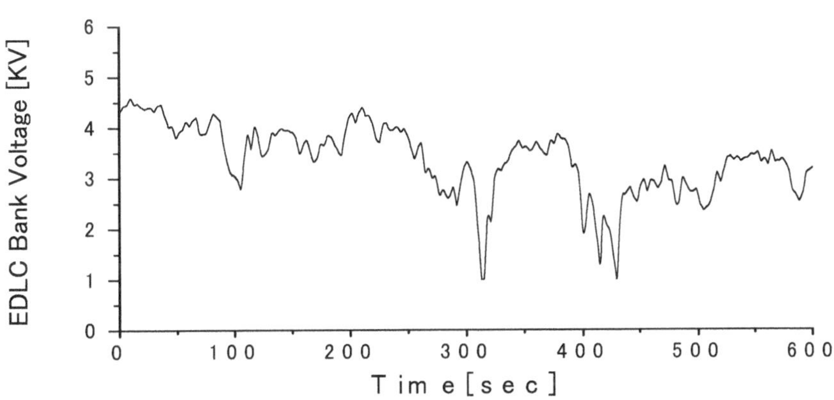

Fig. 7.25 Bank voltage of the EDLC (Case 3)

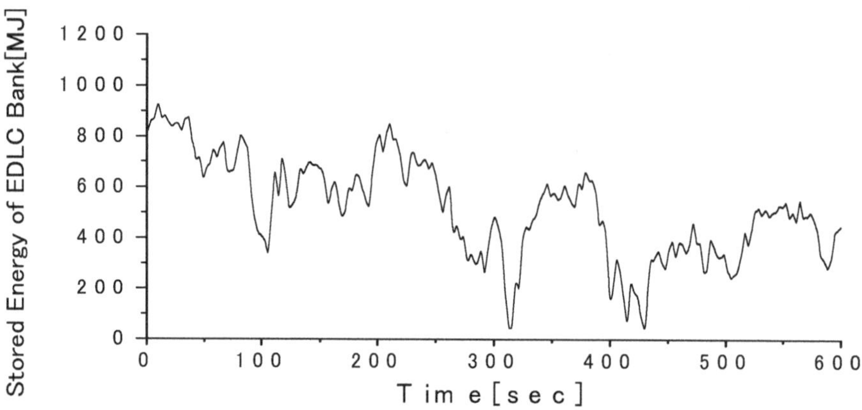

Fig. 7.26 Stored energy of the EDLC bank (Case 3)

7.4 Transient Stability Enhancement of a WTGS by an ECS

Besides wind farm output power smoothing, an ECS can be applied to load leveling, peak saving, sub-synchronous oscillations, and transient and dynamic stability enhancement of a power system. According to the wind farm grid code [8,9], if the voltage of a wind farm remains at a level greater than 15 % of the nominal voltage for a period that does not exceed 0.625 seconds, the plant must stay online. Further, if the voltage does not fall below the minimum voltage indicated by the solid line in Fig. 7.27 and returns to 90 % of the nominal voltage within 3 seconds after the beginning of the voltage drop, the plant must stay online. This study is proposing a new system to achieve the above low voltage ride through requirement for a wind farm during a network disturbance in the power system. Moreover, the transient stability enhancement of the power system including the wind farms is analyzed.

7.4.1 Model System for Transient Analysis

Figure 7.28a shows a model system used in the simulation analyses of the LVRT requirement for a wind generator, where one synchronous generator (SG) is connected to an infinite bus through a transformer and a double circuit transmission line. One aggregate WTGS (IG in Fig. 7.28a) is connected to the network via a transformer and a transmission line. This is called model system I. In an aggregated model, it is assumed that several WTGSs are lumped together to obtain a large WTGS. For wind farm analysis, the aggregated WTGS is replaced by five 10

MW aggregate induction generators, as shown in Fig. 7.28b. This is called model system II. The underground cable of each wind generator is not included in the simulation for ease of simulation. Usually, these lines are not so long for an on-shore wind farm, and the effect may be neglected.

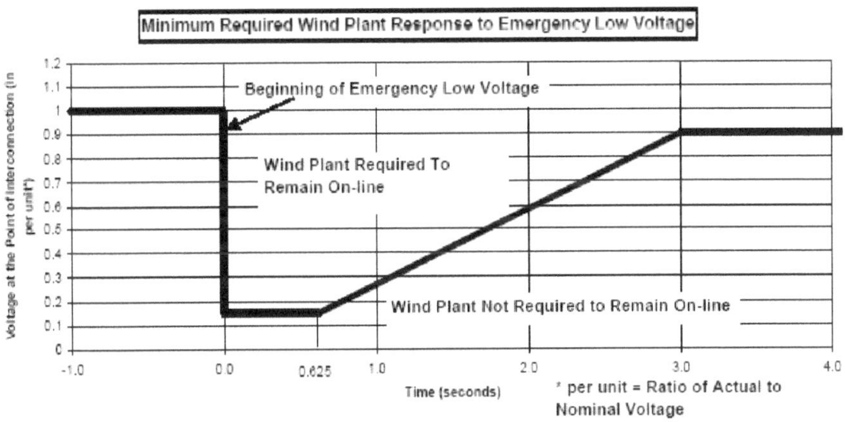

Fig. 7.27 Low voltage ride through standard set by FERC, U.S.[8]

A capacitor bank, C, is used for reactive power compensation of the induction generator at steady state, as described in Chap. 2. The ECS is connected to point K as shown in Figs. 7.28a and b. The AVR (automatic voltage regulator) and GOV (governor) control system models for the synchronous generator are taken from Chap. 2. Generator parameters are shown in Table 7.8. The system base is 100 MVA. The initial values used in the simulation are shown in Tables 7.9 and 7.10 for model systems I and II, respectively. For model systems I and II, the initial values are shown at 0 sec and 100 sec, respectively, just before the occurrence of a network fault.

The ECS modeling and control strategy are the same as that used in Sect. 7.1 of this chapter. The parameters of the PI controller used in Fig. 7.3 are shown in Table 7.11.

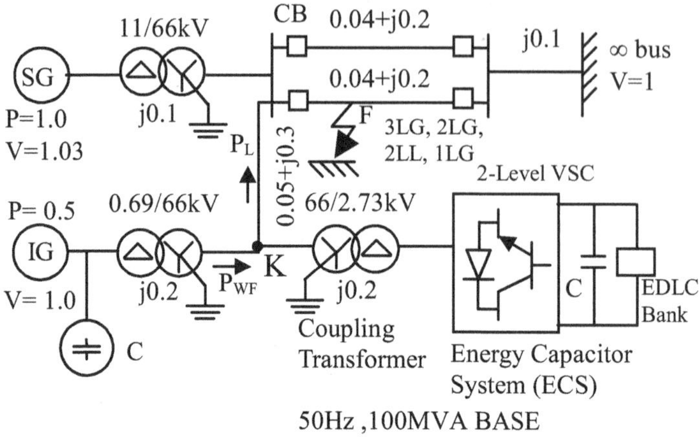

(a) Model System-I

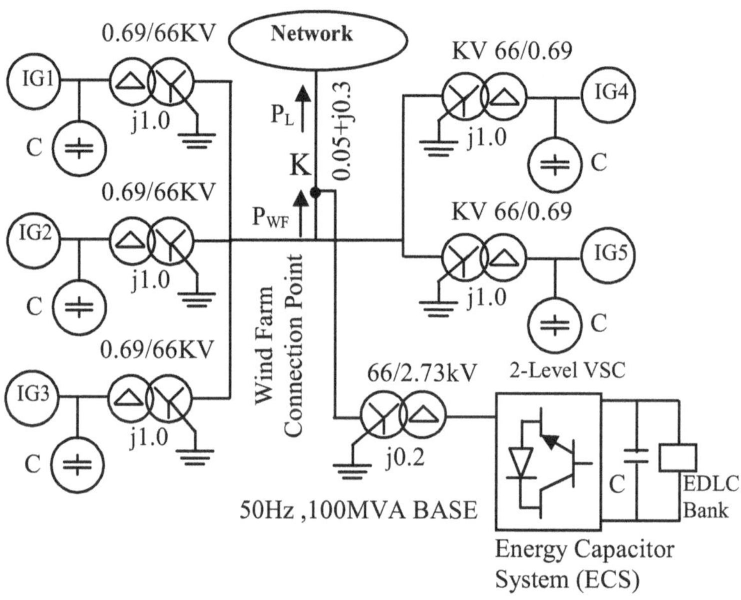

(b) Model system-II

Fig. 7.28 Model systems for transient stability analysis

Table 7.8 Generator parameters

SG		IG	
MVA	100	MVA	50/10
ra (pu)	0.003	r1 (pu)	0.01
xa (pu)	0.13	x1 (pu)	0.1
Xd (pu)	1.2	Xmu (pu)	3.5
Xq (pu)	0.7	r21 (pu)	0.035
Xd' (pu)	0.3	x21 (pu)	0.030
Xq' (pu)	0.22	r22 (pu)	0.014
Xd'' (pu)	0.22	x22 (pu)	0.098
Xq'' (pu)	0.25	H_{WT} (pu)	3.0
Tdo' (sec)	5.0	H_G (pu)	0.3
Tdo'' (sec)	0.04	K_W (pu)	90.0
Tqo'' (sec)	0.05		
H (sec)	2.5		

Table 7.9 Initial values of generators and turbines (model I)

	SG	IG
P(pu)	1.0	0.50
V(pu)	1.03	0.999
Q(pu)	0.334	0.000 (0.239)*
Efd(pu)	1.803	-
Tm(pu)	1.003	-
δ (deg)	50.72	-
slip	0.0	–1.09%
Vw (m/s)	-	11.797
β (deg)	-	0

* Reactive power drawn by an induction generator

Table 7.10 Initial values of generators and turbines (model II)

	SG	IG1	IG2	IG3	IG4	IG5
P(pu)	1.0	0.098	0.10	0.10	0.097	0.098
V(pu)	1.03	1.002	1.001	1.001	1.002	1.002
Q(pu)	0.331	0.001 (0.046)*	0.000 (0.047)*	0.000 (0.047)*	0.001 (0.046)*	0.001 (0.046)*
Efd(pu)	1.80	-	-	-	-	-
Tm(pu)	1.003	-	-	-	-	-
δ (deg)	50.75	-	-	-	-	-
slip	0.0	−1.04%	−1.15%	−1.13%	−1.04%	−1.04%
Vw (m/s)	-	11.67	12.59	12.08	11.65	11.66
β (deg)	-	0	6.48	2.54	0	0

* Reactive power drawn by an induction generator

Table 7.11 The parameters of the PI controllers used in Sect. 7.4

	PI-1	PI-2	PI-3	PI-4
K_p	3.0	2.0	3.0	0.03
T_i	0.1	0.004	0.1	0.002

7.4.2 Simulation Results of Transient Analysis

In this study, the simulation results are described, only in the light of the US grid code set by the Federal Energy Regulatory Commission (FERC) [8]. This book is proposing an ECS with a suitable control strategy to enhance the LVRT capability of a fixed speed wind generator under network disturbances.

When a network fault occurs, the reactive power demand of the wind farm is supplied according to the error signal between the wind farm terminal voltage, Vk, and the reference voltage.

On the other hand, the ECS is forced to work only to store transient energy by switching off the switch, g2, of the DC-DC buck/boost converter. Therefore, the active power can be controlled and this would be effective in enhancing the transient stability of the rest of the system.

To obtain realistic responses, the two-mass shaft model of a WTGS is considered. All types of damping are disregarded to obtain the worst-case scenario. A symmetrical three-line-to-ground fault, 3LG, and unsymmetrical double-line-to-ground fault, 2LG (phases B, C, and ground), a double-line fault, 2LS (between phases B and C), and a single-line-to-ground fault, 1LG (phase C and ground) are considered as the network disturbances, which occur at fault point F in Fig. 7.28.

Simulations have been performed by using PSCAD/EMTDC, which uses a fixed time step algorithm. The simulation time step chosen is 0.01 msec. To verify the effectiveness of the control strategy of the ECS for achieving the LVRT requirement, three cases are considered as explained below.

Case 1: In this case, the aggregated model of the wind farm shown in Fig. 7.28a is considered, where one large wind generator represents several wind generators. It is assumed in the simulation that wind speed is constant and equivalent to the rated speed of 11.8 m/s. Because it may be considered that wind speed does not change dramatically during the short time interval of the simulation. The pitch controller is not considered in this case to demonstrate the effectiveness of the proposed ECS for achieving the LVRT requirement. The simulation time duration is 4.0 sec. A fault occurs at 0.1 sec at fault point F in Fig. 7.28a, and then the circuit breakers (CB) on the faulted lines are opened at 0.2 sec, i.e., the fault is cleared within the permissible range of the grid code [139]. Finally, at 1.0 sec the circuit breakers are reclosed.

The response of the induction generator terminal voltage is shown in Fig. 7.29 with and without an ECS, when a severe 3LG fault occurs in the model system. In the case without an ECS, the voltage drop occurs at the terminal of the induction generator, as shown in the figure. Therefore, the electromagnetic torque of the induction generator also drops suddenly because the electromagnetic torque is proportional to the square of the terminal voltage. But the mechanical torque of the wind turbine doesn't change rapidly during that short time interval. As a result, the turbine hub and generator rotor accelerate due to the large difference between the mechanical and electromagnetic torques of the WTGS, as shown in Fig. 7.30. But when the ECS is used, the necessary reactive power is supplied from the ECS properly according to the error signal between the wind farm terminal and its reference, so that the terminal voltage of the wind generator can be returned to the pre-fault level. Thus the electromagnetic torque can be restored quickly, and the WGTS becomes stable with an ECS. From Fig. 7.29, it can be seen clearly that an ECS can enhance the low voltage ride through capability of the wind generator under the severe 3LG fault. Moreover, the ECS absorbs the transient energy, which enhances the transient stability of the SG, as shown in Fig. 7.31. Figure 7.32 shows the active and reactive power responses of the ECS. The responses of the DC-link capacitor voltage, the EDLC bank voltage, and the stored energy of the EDLC bank are shown in Figs. 7.33 – 7.35, respectively.

Figures 7.36 – 7.39 show simulation results for a 2LG fault. Figure 7.36 shows that an ECS can enhance the LVRT capability of the wind generator during a 2LG fault. But without the ECS, the LVRT requirement of the wind generator cannot be achieved. The responses of the turbine hub and IG rotor speed, and the real and reactive power of the ECS are shown in Figs. 7.37 and 7.38, respectively. The load angle of the synchronous generator is shown in Fig. 7.39.

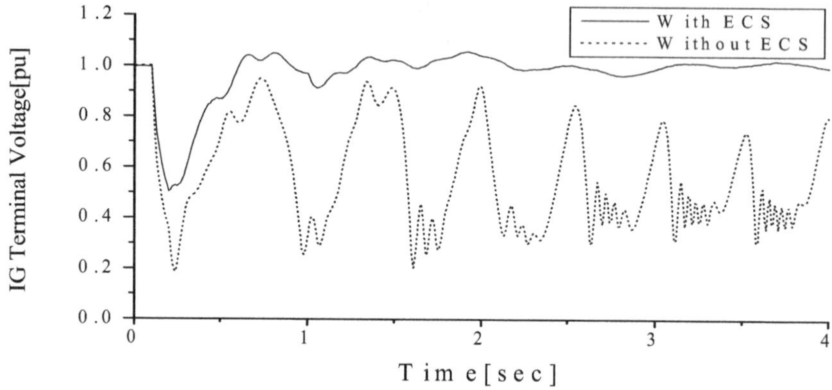

Fig. 7.29 IG terminal voltage (Case 1, 3LG fault)

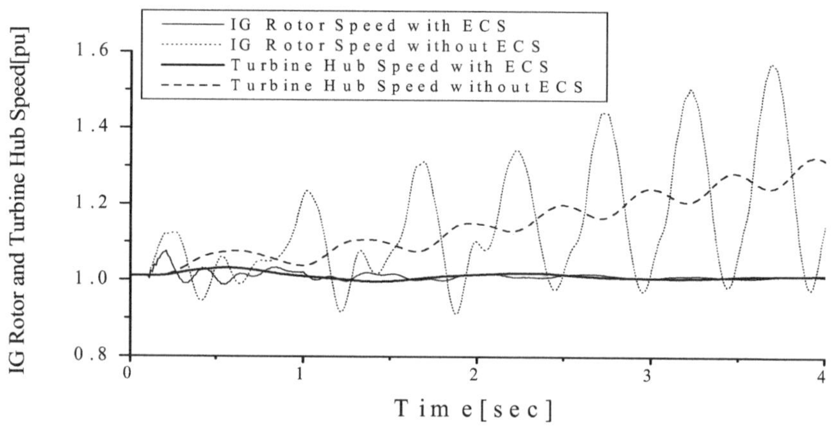

Fig. 7.30 Turbine hub and IG rotor speeds (Case 1, 3LG fault)

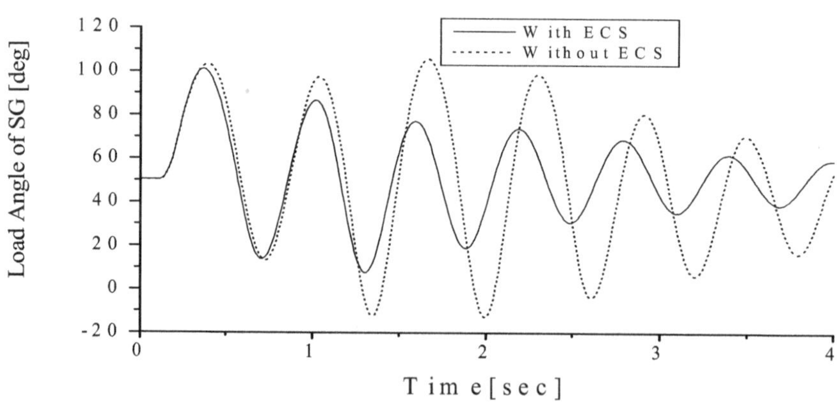

Fig. 7.31 Load angle of the SG (Case 1, 3LG fault)

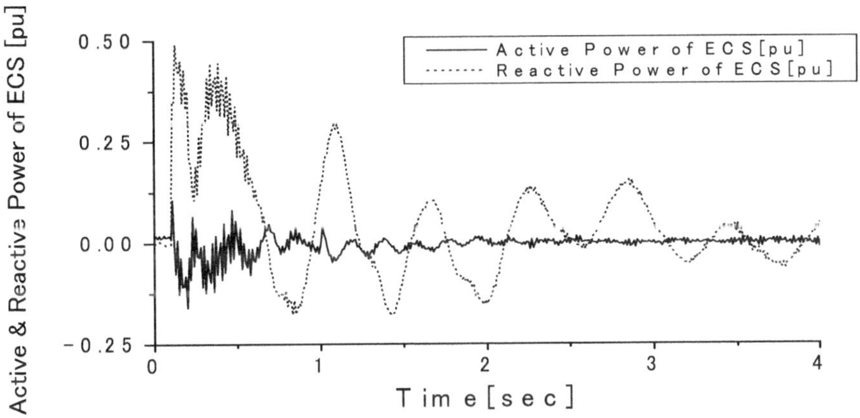

Fig. 7.32 Active and reactive power of the ECS (Case 1, 3LG fault)

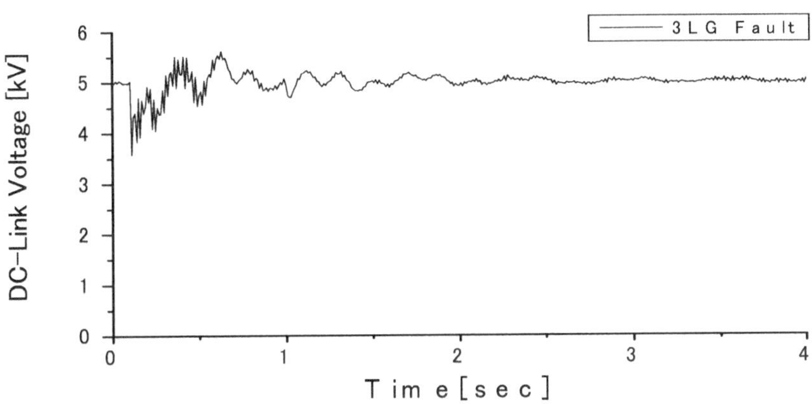

Fig. 7.33 DC-link voltage of the ECS (Case 1, 3LG fault)

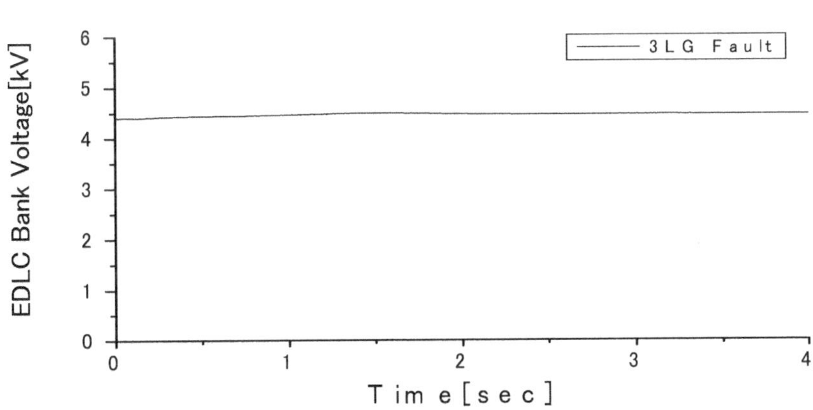

Fig. 7.34 EDLC bank voltage (Case 1, 3LG fault)

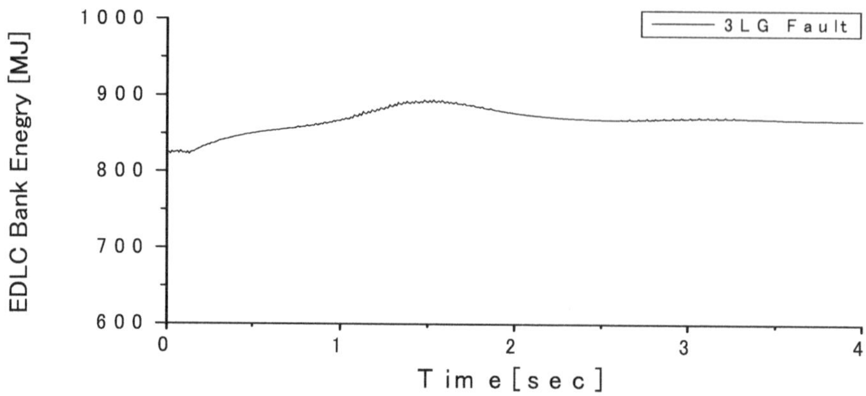

Fig. 7.35 EDLC stored energy (Case 1, 3LG fault)

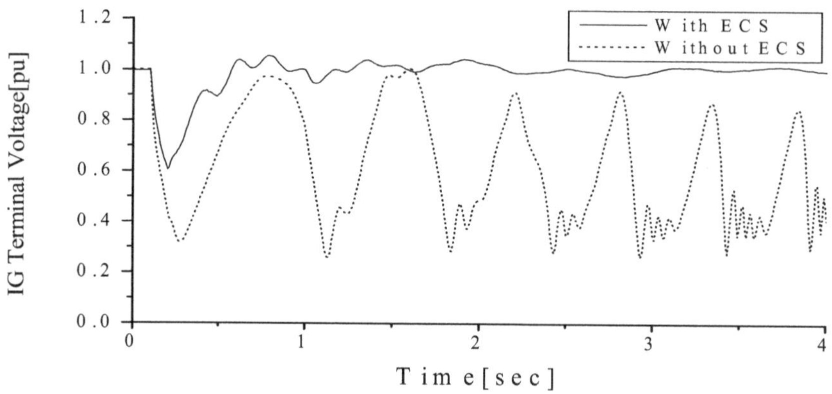

Fig. 7.36 IG terminal voltage (Case 1, 2LG fault)

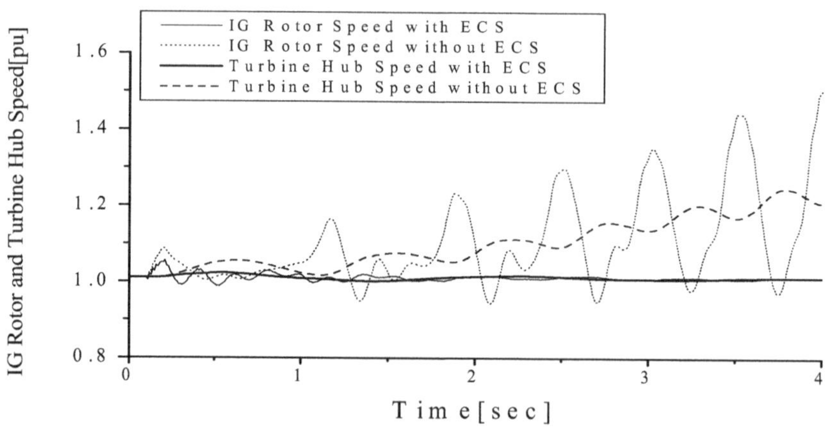

Fig. 7.37 Turbine hub and IG rotor speeds (Case 1, 2LG fault)

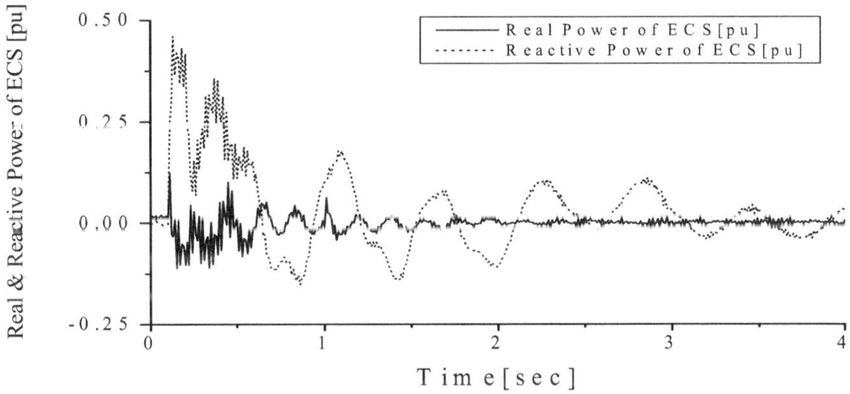

Fig. 7.38 Real and reactive power of ECS (Case 1, 2LG fault)

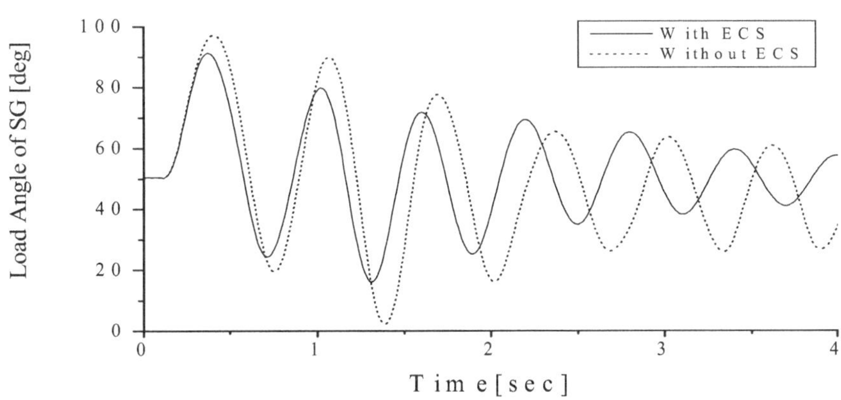

Fig. 7.39 Load angle of the SG (Case1, 2LG fault)

Figures 7.40 – 7.42 show simulation results for a 2LS fault. During the 2LS fault, the ECS can return the terminal voltage of the wind generator to pre-fault level faster than without an ECS, as shown in Fig. 7.40. The turbine hub and wind generator rotor speeds are shown in Fig. 7.41. It is seen that an ECS can stabilize the WTGS more quickly than that without ECS. The load angle response of the SG with and without an ECS is shown in Fig. 7.42.

The response of the wind generator terminal voltage with and without an ECS during the 1LG fault is shown in Fig. 7.43, from which it is also seen that an ECS can enhance the stability of the wind generator.

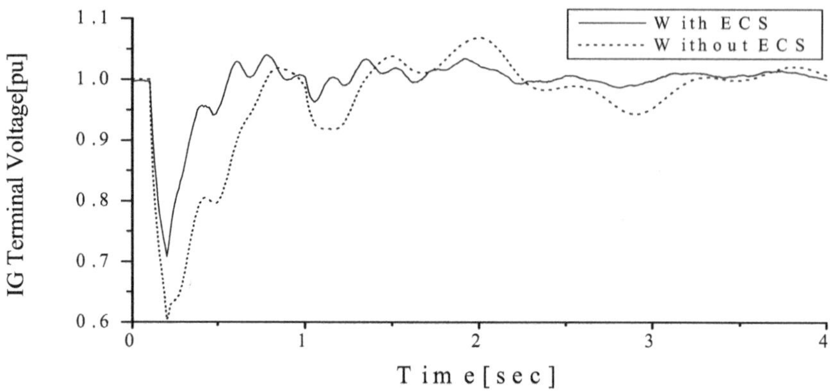

Fig. 7.40 IG terminal voltage (Case 1, 2LS fault)

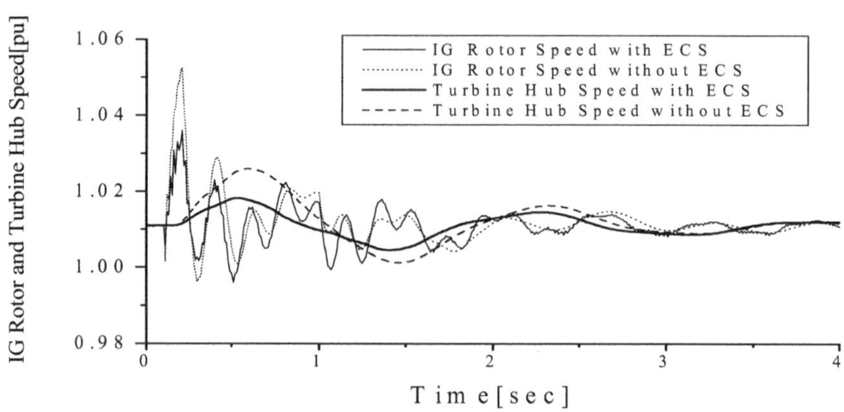

Fig. 7.41 Turbine hub and IG rotor speed (Case 1, 2LS fault)

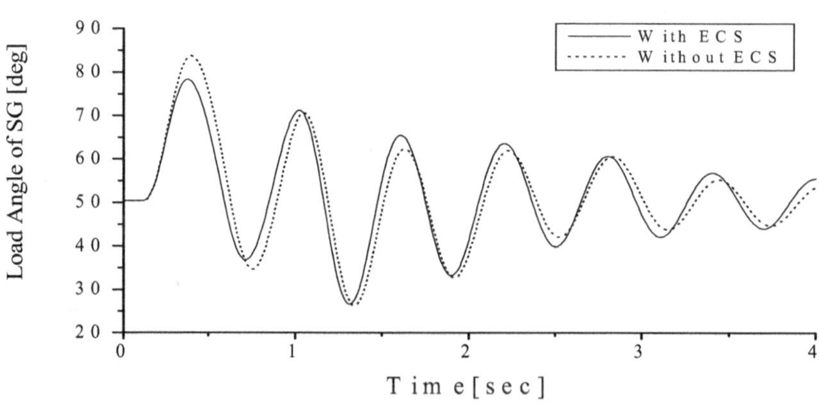

Fig. 7.42 Load angle of the SG (Case 1, 2LS fault)

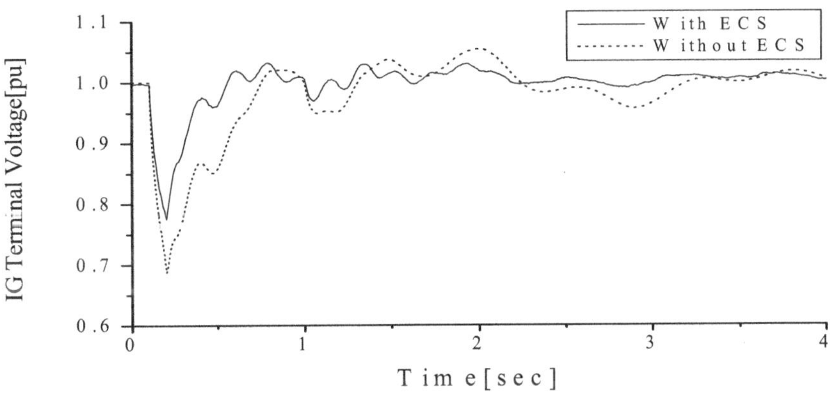

Fig. 7.43 IG terminal voltage (Case 1, 1LG fault)

Case 2: In this case, the permanent fault due to unsuccessful reclosing of the circuit breakers is analyzed. The circuit breakers are usually reclosed automatically to improve service continuity. The re-closure may be either high-speed or with a time delay. High-speed re-closure refers to the closing of circuit breakers after a time just long enough to permit fault-arc de-ionization. However, high-speed re-closure is not always acceptable. Reclosure into a permanent fault, i.e., unsuccessful reclosure may cause system instability. Thus, the application of automatic reclosing is usually constrained by the possibility of a persistent fault, which would create a second fault after reclosure. It is reported herein that an ECS can enhance the transient stability of the synchronous generator during the permanent fault condition.

In this case, the transient stability analysis is carried out when the wind speed is at the rated level of 11.8 m/sec. In this case, the pitch controller is also not considered. Model system I shown in Fig. 7.28a is considered. It is considered that a 3LG fault occurs at 0.1 sec, circuit breakers on the faulted line are opened at 0.2 sec, and are closed again at 1.0 sec. Because the reclosing of the circuit breakers is considered unsuccessful due to a permanent fault, the circuit breakers are re-opened at 1.1 sec. It is assumed that the circuit breaker clears the line when the current through it crosses the zero level. The simulation time duration is 10.0 sec.

Figure 7.44 shows the responses of the wind turbine and induction generator rotor speeds. It is seen that a WTGS becomes unstable when an ECS is not considered. But with an ECS, the WTGS becomes stable. The IG terminal voltage can return its pre-fault level when an ECS is used, as shown in Fig. 7.45, i.e., the LVRT requirement for a WTGS is achieved even in the case of the permanent fault due to the unsuccessful reclosing. Figure 7.46 shows the responses of the synchronous generator load angle with and without an ECS. It is clearly seen that the synchronous generator is transiently stable well when an ECS is used. The size ratio of the ECS allows it to influence the stability of the SG. This fact also indicates that an ECS can stabilize well the entire power system.

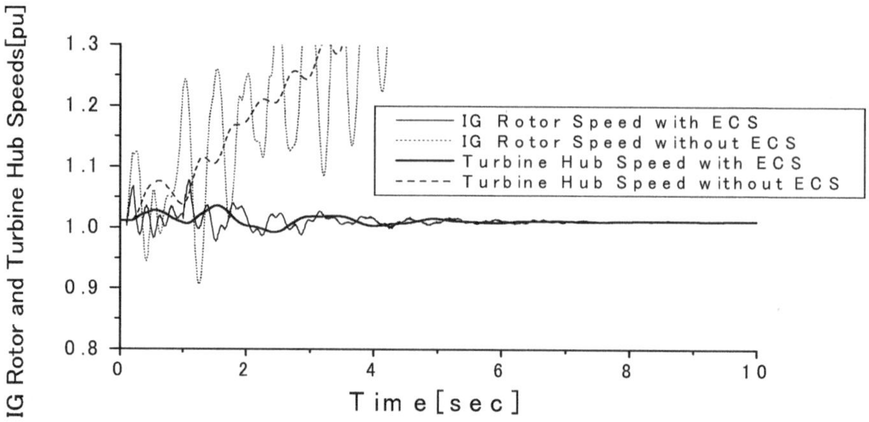

Fig. 7.44 Turbine hub and IG rotor speeds (Case 2, 3LG permanent fault)

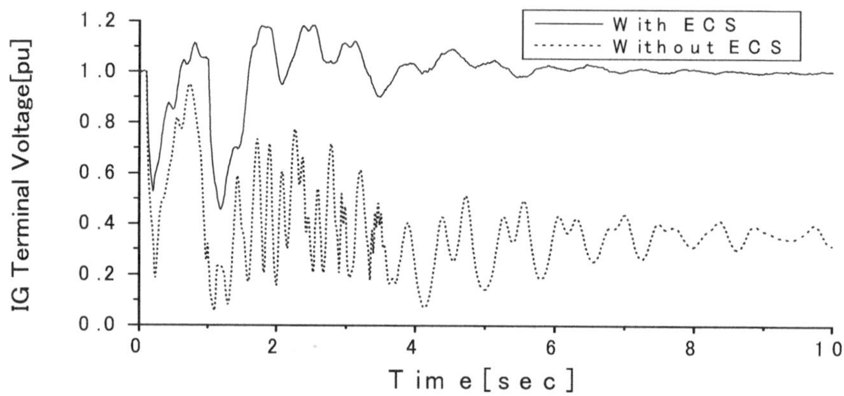

Fig. 7.45 IG terminal voltage (Case 2, 3LG permanent fault)

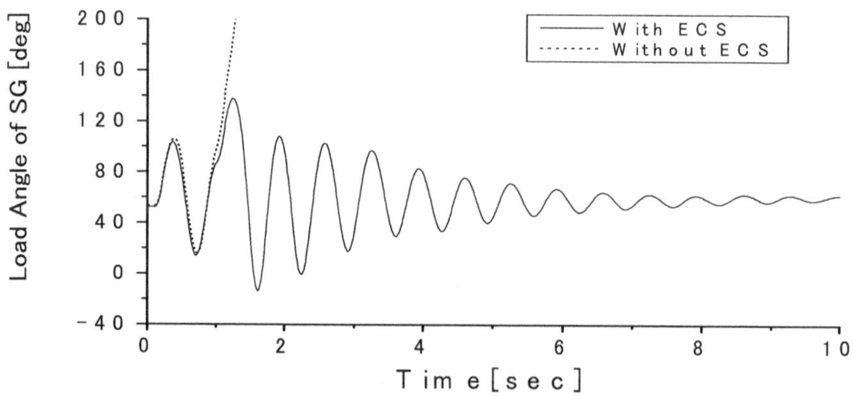

Fig. 7.46 Load angle of the SG (Case 2, 3LG permanent fault)

Case 3: In this case, another wind farm model with five wind generators shown in Fig. 7.28b is considered. Real wind speed data shown in Fig. 7.47, which were obtained on Hokkaido Island, Japan, are used at each wind generator. The data were measured at a single location using an isolated type of wind turbine. A pitch controller is used in this case to maintain the output power at the rated level when the wind speed is over the rated speed. In this case, a fault occurs at 100.1 sec, the circuit breakers (CB) on the faulted lines are opened at 100.2 sec, and at 101.0 sec are reclosed. The simulation time duration is 4.0 sec. It is seen from Fig. 7.48 that the voltage at the high-voltage (HV) side of the wind farm substation transformer

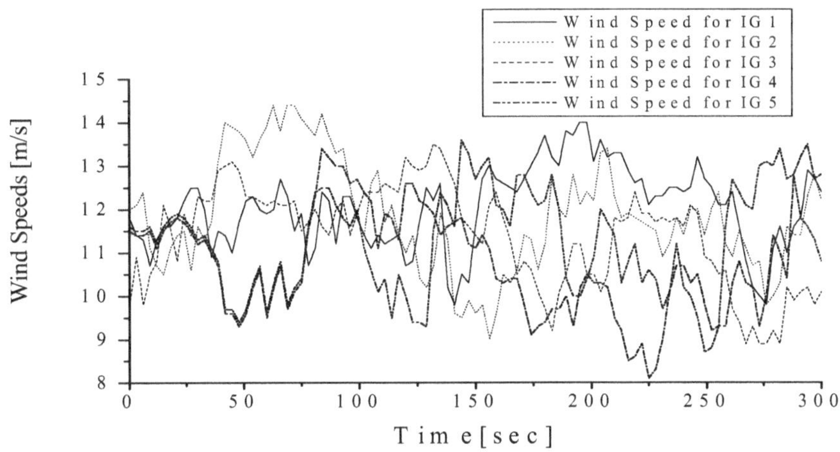

Fig. 7.47 Wind speed data for five IGs (Case 3, 3LG fault)

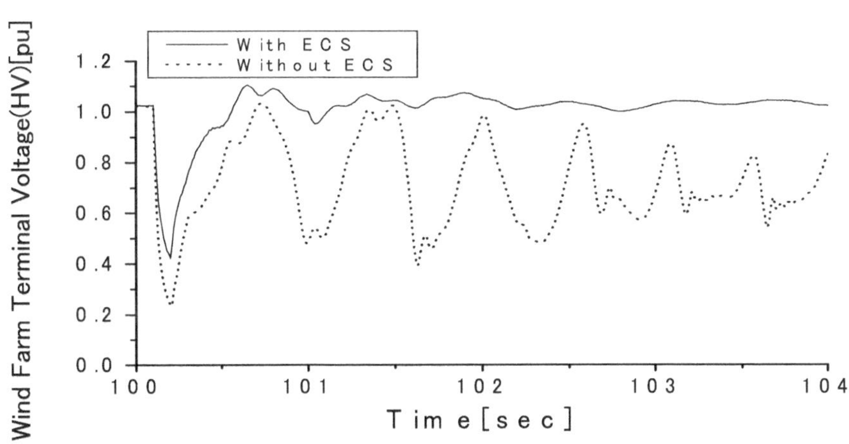

Fig. 7.48 Wind farm connection point voltage (Case 3, 3LG fault)

failed to return to 90 % of the rated line voltage during the severe 3LG fault. But with an ECS, the wind farm connection point voltage can achieve the requirement of the U.S. grid code mentioned at the beginning of this section.

7.5 Chapter Summary

Due to the natural wind speed variation, the output power and terminal voltage of a fixed speed wind farm fluctuate randomly. This chapter proposes a system using an ECS where smoothed line power and constant terminal voltage can be obtained from a fixed speed wind farm, because the ECS has both real and reactive power controllability. The modeling and control strategy for a ECS are presented clearly. The exponential moving average is introduced to calculate the reference of wind farm output power. Additionally, by taking advantage of an ECS, the most economical and performance-effective hydrogen generator topology is integrated at the wind farm terminal. Simulation results validate the cooperative control of the proposed system. It can be concluded that the proposed system composed of a fixed speed wind farm, hydrogen generator, and an ECS can be a good solution to wind power application.

It is also shown that the ECS can enhance the LVRT capability of wind farms according to the grid code. Besides these, the ECS can also enhance the transient stability of power systems including wind farms. The effectiveness of the proposed control system is verified with different types of fault conditions at different locations in the power system model.

Chapter 8

Stability Enhancement of VSWT-PMSG

Until the end of the 1990s, the constant speed concept dominated the market, and it still represents a significant part of the operating wind turbine population. However, recently the variable speed wind turbine generator system (WTGS) is drawing the attention of the power grid operator. The variable speed WTGS uses the power electronic converters that enable decoupling the grid frequency from the real-time rotational frequency, as imposed by the instantaneous wind speed and the wind turbine control system. Variable speed operation enables an optimization of the performance, reduces the mechanical loading, and at the same time delivers various options for active power plant control. Decoupling the electrical and rotor frequencies absorbs wind speed fluctuations, allowing the rotor to act as a (accelerating and decelerating) flywheel, and thus smoothing out spikes in power, voltage, and torque [6].

The doubly fed induction generator (DFIG), wound field synchronous generator (WFSG), and permanent magnet synchronous generator (PMSG) are widely used as variable speed wind generators. In this chapter, the stabilization of the variable speed wind turbine (VSWT) driving a PMSG is analyzed. Two types of VSWT-PMSG topologies are presented with their suitable control strategies. Finally, the transient stabilities of both topologies during both symmetrical and unsymmetrical fault conditions are analyzed.

In the PMSG, the excitation is provided by permanent magnets instead of field windings. Permanent magnet machines are characterized as having large air gaps, which reduce flux linkage even in machines with multi-magnetic poles [101, 102]. As a result, low rotational speed generators can be manufactured in relatively small sizes with respect to their power ratings. Moreover, a gearbox can be omitted due to the low rotational speed in PMSG wind generators, resulting in low cost. In a recent survey, gearbox is found to be the most critical component, since its downtime per failure is high in comparison to other components in WTGS [140]. The wind turbine characteristics used in this analysis are explained in Sect. 2.4.2.

8.1 Maximum Power Point Tracking

For the maximum power point tracking (MPPT) operation, rotor speed is used as a controller input instead of wind speed, as shown in Fig. 8.1, because the rotor speed can be measured more precisely and more easily than the wind speed. The details of MPPT are also available in Sect. 2.4.2.

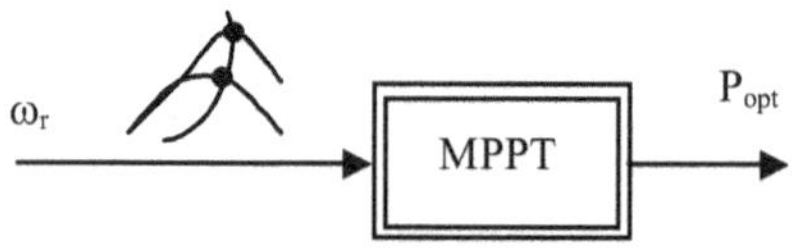

Fig. 8.1 MPPT searching methodology block

The range of rotor speed variation is, in general, approximately 5 to 16 rpm. If the reference optimum power, P_{opt}, is greater than the rated power of the PMSG, then the pitch controller shown in Fig. 8.2 is used to control the rotational speed. Therefore, the reference optimum power will not exceed the rated power of the PMSG. The pitch servo is modeled with a first order delay system with a time constant, T_d, of 3.0 sec. Because the pitch actuation system cannot, in general, respond instantly, a rate limiter with the value of 10°/s is added.

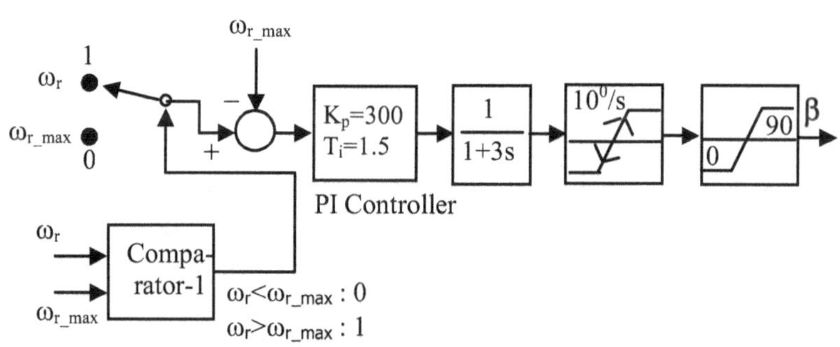

Fig. 8.2 Pitch controller

8.2 Modeling of a PMSG

In the simulation analyses, the PMSG model available in the package software PSCAD/EMTDC is used[1]. The nominal speed is considered as the maximum rotor speed, ω_{r_max}. The pitch controller is activated when the rotor speed exceeds the maximum rotor speed.

8.3 VSWT-PMSG with Converter-DC Link-Inverter Topology

In this section, the direct drive VSWT-PMSG concept is adopted with the use of a fully controlled frequency converter. The frequency converter consists of a generator side AC/DC converter, a DC link capacitor, and a grid side DC/AC inverter. Each of converter/inverter is a standard three-phase two-level unit, composed of six IGBTs and antiparallel diodes. The electrical scheme of the VSWT-PMSG topology is shown in Fig. 8.3. The control strategy of each converter is shown below.

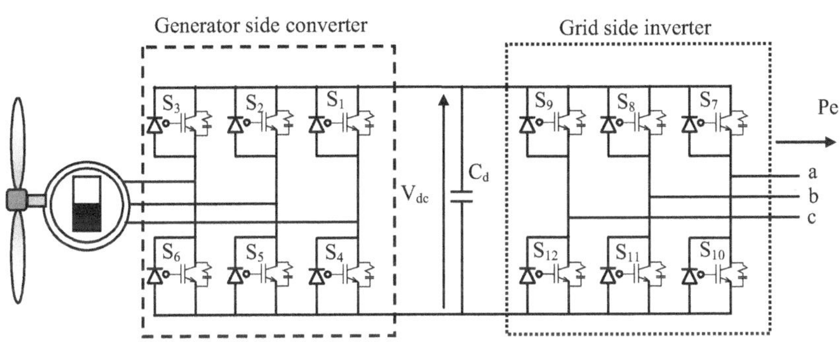

Fig. 8.3 Electrical scheme (1) of a VSWT-PMSG

8.3.1 Modeling and Control Strategy of Generator Side Converter

The well-known cascaded control scheme shown in Fig. 8.4 is used as the control methodology for the generator side converter. Because this converter is directly connected to the PMSG, its q-axis current can control the active power. The active power reference, P_{opt}, is determined to provide maximum power to the grid. On the other hand, the d-axis stator current can control the reactive power. The reac-

[1] For the latest information on PSCAD/EMTDC, visit at http://pscad.com

tive power reference is set to zero for unity power factor operation. The angle, θ_r, for the transformation between abc and dq variables is calculated from the rotor speed of the PMSG.

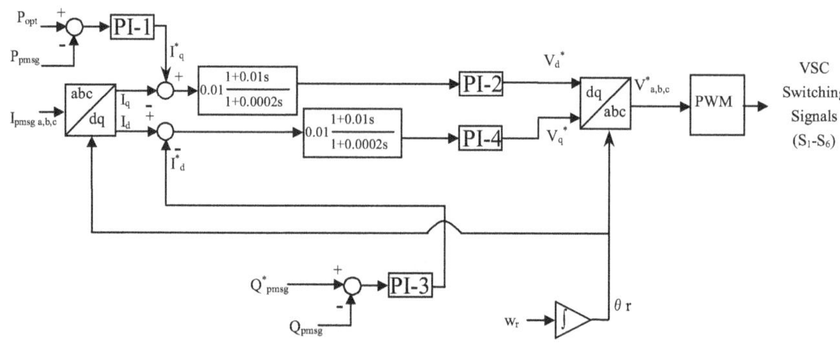

Fig. 8.4 Control block diagram of the generator-side converter

8.3.2 Modeling and Control Strategy of Grid Side Inverter

Control blocks for the grid side inverter are shown in Fig. 8.5 which is based on the cascaded control scheme. The modeling of the grid side inverter is described in Sect. 4.2 of Chap. 4. The dq quantities and three-phase electrical quantities are related to each other by a reference frame transformation. The angle of the transformation is detected from the three phase voltages (v_a,v_b,v_c) at the high-voltage side of the grid side transformer. The DC voltage of the DC-link capacitor is controlled constant by two PI controllers. The d-axis current can control the DC-link voltage. On the other hand, the q-axis current can control the reactive power of the grid side inverter. The reactive power reference is set that the terminal voltage at the high-voltage side of the transformer remains constant. Therefore, three PI controllers are used to control the reactive power of the grid side inverter. The additional PI controller provides excellent transient characteristics during network disturbance, as shown later.

In both converter and inverter, the triangular carrier signal is used as the carrier wave of PWM operation. The carrier frequency chosen is 1000 Hz for the converter and 1050 Hz for the inverter, respectively. The DC-link capacitor value chosen is 10000 μF. The rated DC-link voltage is 2.3 kV.

8.3.3 Model System Used in Sect. 8.3

The model system used for the transient stability analysis of the VSWG-PMSG is shown in Fig. 8.6. Here, one PMSG is connected to an infinite bus through the generator side converter, DC-link capacitor, grid side inverter, transformer, and double circuit transmission line. The parameters of the PMSG are shown in Table 8.1. The system base is 5 MVA.

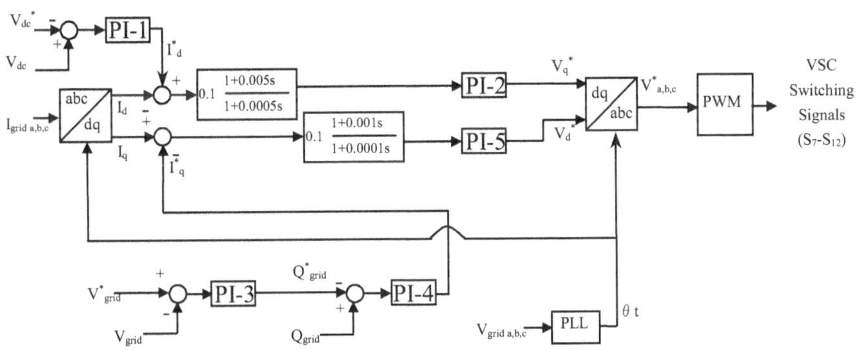

Fig. 8.5 Control block diagram of the grid-side inverter

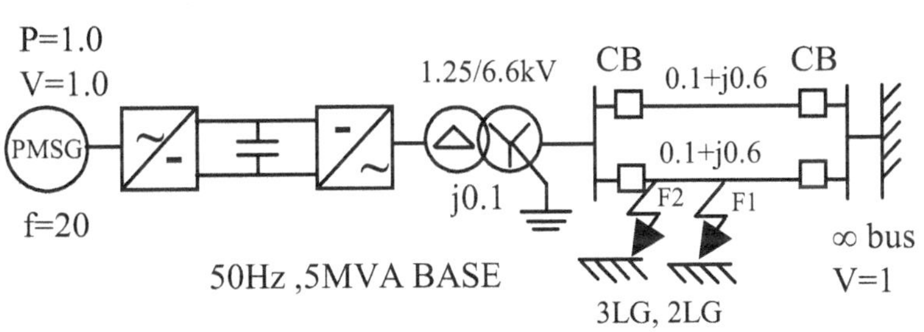

Fig. 8.6 Model system used in Sect. 6.3

Table 8.1 Generator parameters

Rated power	5 [MW]	Stator resistance	0.01[pu]
Rated voltage	1.0 [kV]	d-axis reactance	1.0 [pu]
Frequency	20 [Hz]	q-axis reactance	0.7 [pu]
Number of poles	150	Field flux	1.4 [pu]
H	3.0 [sec]		

8.3.4 Simulation Analysis

A symmetrical three-line-to-ground fault, 3LG, and an unsymmetrical double-line-to-ground fault, 2LG (phases B, C, and ground) are considered network disturbances each occurs at different fault points of the transmission line, as shown in Fig. 8.6. The fault occurs at 0.1 sec, the circuit breakers (CB) on the faulted lines are opened at 0.2 sec, and at 1.0 sec the circuit breakers are re-closed. In the transient stability analysis, the wind speed is kept constant at the rated speed at which the PMSG reference power is at the rated level, assuming that the wind speed doesn't change dramatically within this small time duration. The time step and simulation time have been chosen as 0.00001 sec and 10 sec, respectively. Simulations were done by using PSCAD/EMTDC[2] [126]. For detailed transient stability analysis of VSTW-PMSG, three cases are considered as described below. The generator side converter and grid side inverter parameters are shown in Tables 8.2 and 8.3, respectively.

Case 1: In this case, a 3LG fault is considered to occur at the middle of one transmission line (fault point F1) of Fig. 8.6. The grid side inverter can provide the necessary reactive power during the network disturbance, as shown in Fig. 8.7. Therefore, the terminal voltage can return to its pre-fault level, as shown in Fig. 8.8. The response of the PMSG rotor speed is shown in Fig. 8.9. Depending on the rotor speed the reference of the generator side converter is determined, as shown in Fig. 8.10. The real power response of the grid is shown in Fig. 8.11. The pitch controller is activated when the rotor speed exceeds the nominal speed of the PMSG. The turbine blade pitch angle is shown in Fig. 8.12. The rotor of the PMSG needs some time to reach steady state due to the slow response of the pitch controller servo system. The response of the DC-link voltage is shown in Fig. 8.13. From the simulation results, it is seen that the proposed control system can enhance the transient stability of the VSWT-PMSG when a 3LG fault occurs far from the wind generator.

Table 8.2 PI controller parameters of the generator side converter shown in Sect. 8.3.1

	PI-1	**PI-2**	**PI-3**	**PI-4**
K_p	0.2	1.0	0.2	1.0
T_i	0.2	0.025	0.2	0.025

[2] For the latest information on PSCAD/EMTDC, visit at http://pscad.com

Table 8.3 PI controller parameters of the grid side inverter shown in Sect. 8.3.2

	PI-1	PI-2	PI-3	PI-4	PI-5
K_p	1.0	0.5	3.0	1.0	0.1
T_i	0.5	0.008	0.8	0.5	0.008

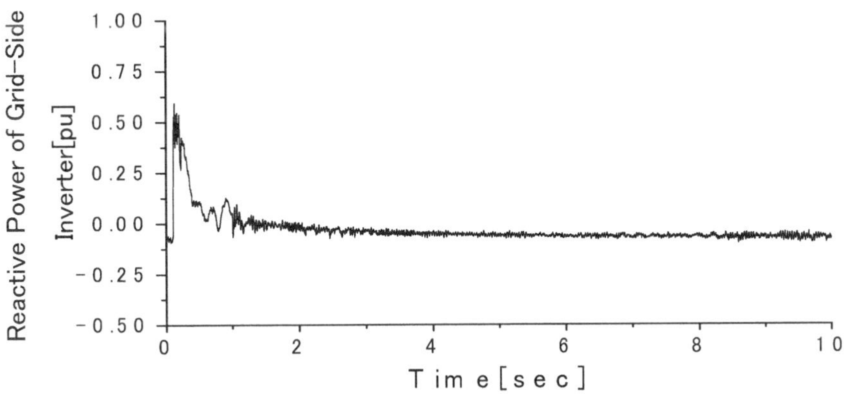

Fig. 8.7 Reactive power of the grid side inverter (Case 1)

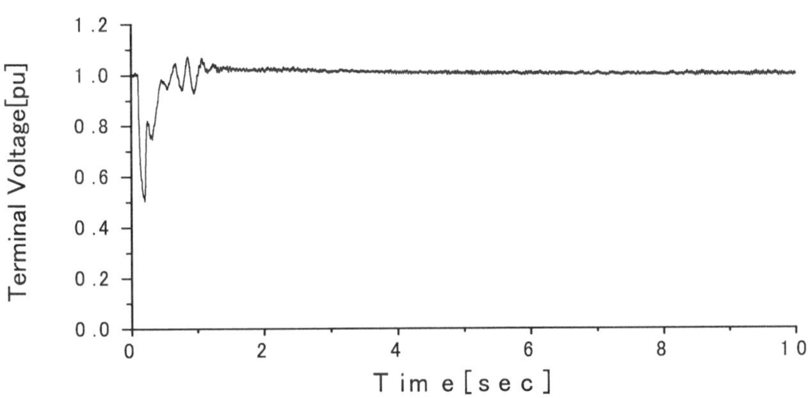

Fig. 8.8 Terminal voltage of the grid (Case 1)

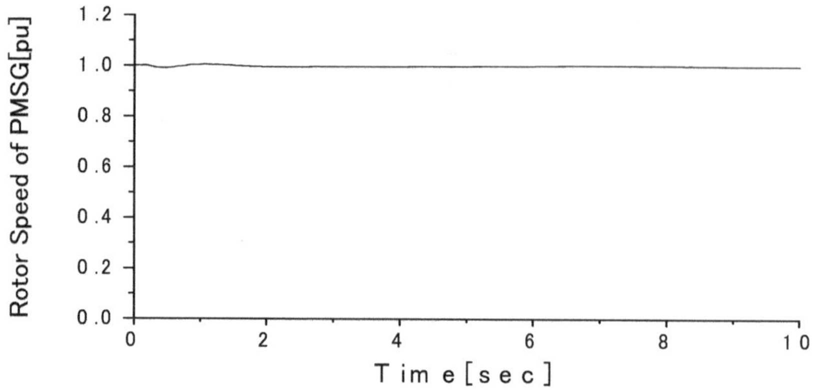

Fig. 8.9 Rotor speed of the PMSG (Case 1)

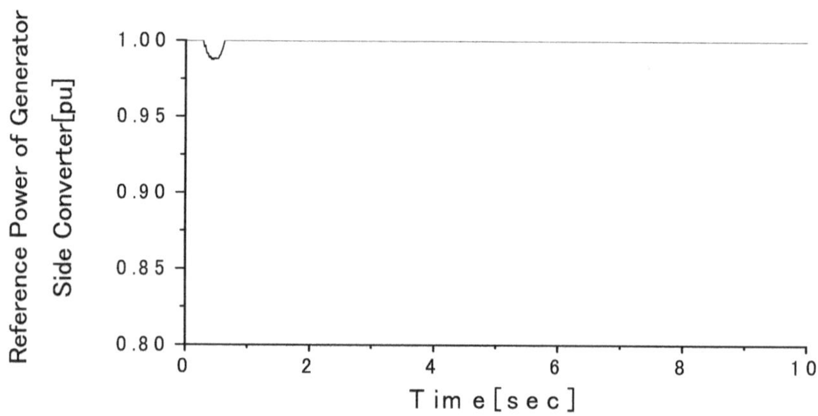

Fig. 8.10 Real power reference of the generator side converter (Case 1)

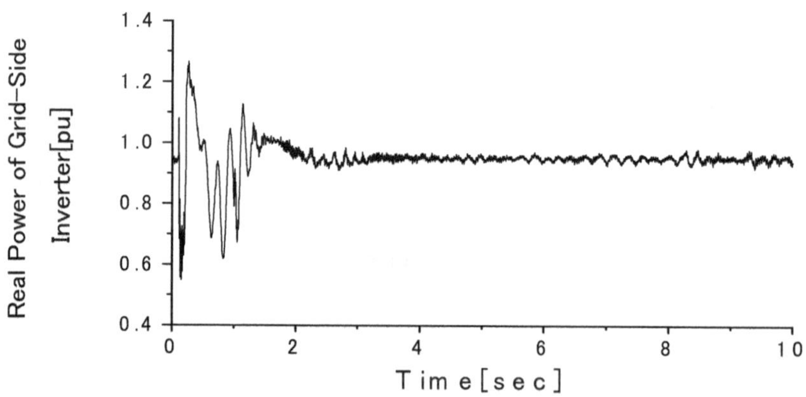

Fig. 8.11 Real power of the grid side inverter (Case 1)

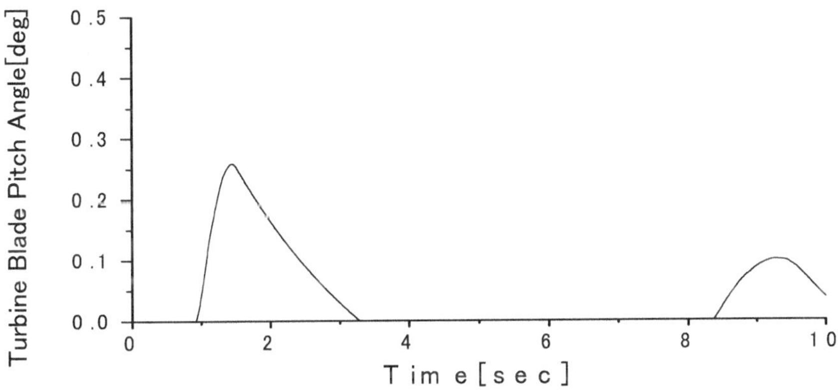

Fig. 8.12 Turbine blade pitch angle (Case 1)

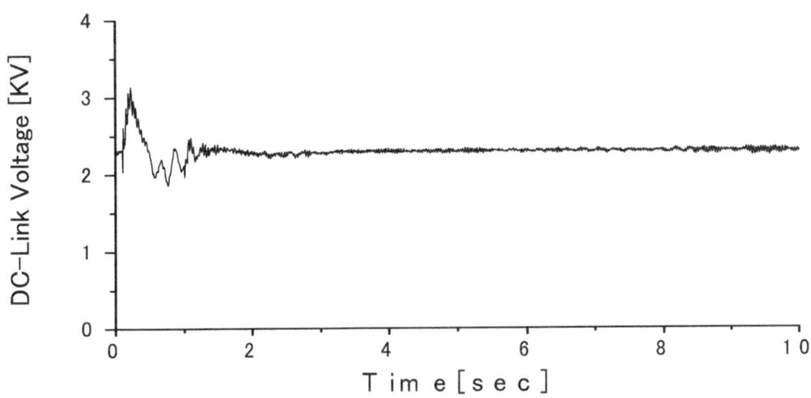

Fig. 8.13 DC-link circuit voltage (Case 1)

Case 2: In this case, a 3LG fault is considered to occur at the sending end of one transmission line (fault point F2) of Fig. 8.6. The response of the grid side reactive power, terminal voltage of the grid, real power, rotor speed of the PMSG, turbine blade pitch angle, and DC-link voltage are shown in Figs. 8.14 – 8.19, respectively. From these results, it is seen that the proposed control system can also enhance the transient stability of a VSWT-PMSG when a 3LG fault occurs close to the wind generator.

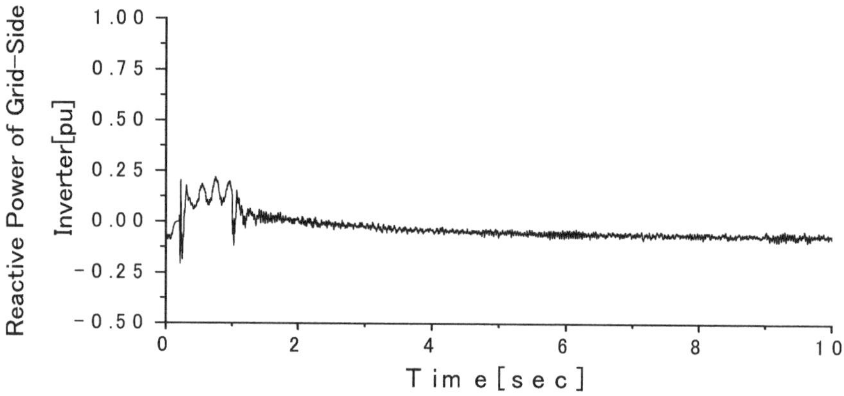

Fig. 8.14 Reactive power of the grid side inverter (Case 2)

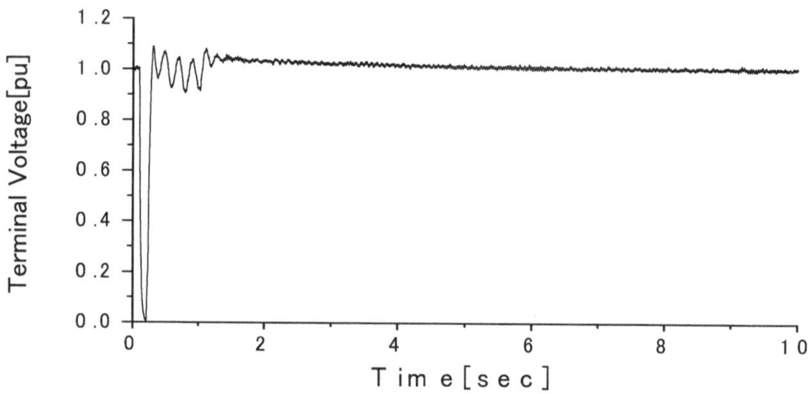

Fig. 8.15 Terminal voltage of the grid (Case 2)

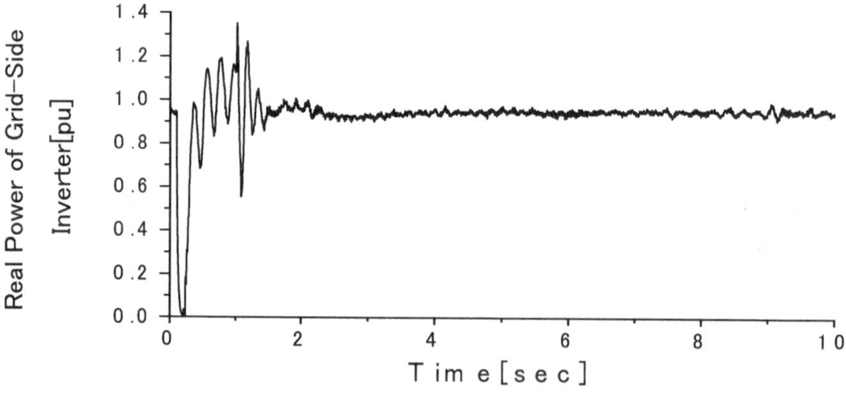

Fig. 8.16 Real power of the grid side inverter (Case 2)

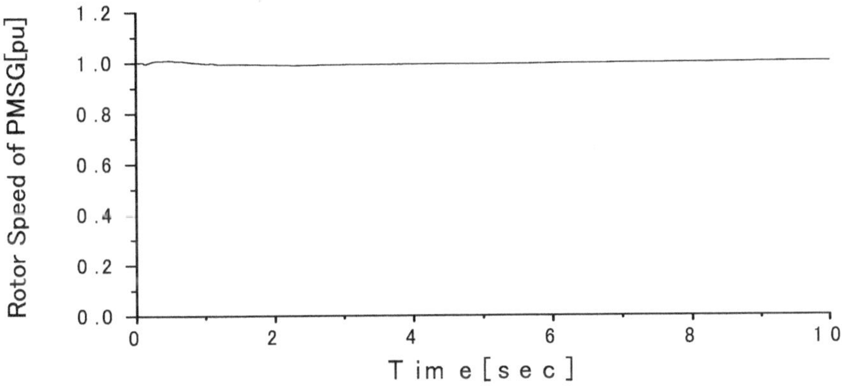

Fig. 8.17 Rotor speed of the PMSG (Case 2)

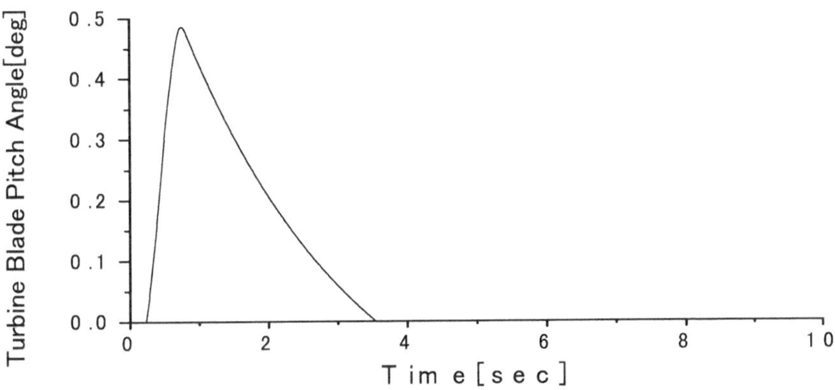

Fig. 8.18 Turbine blade pitch angle (Case 2)

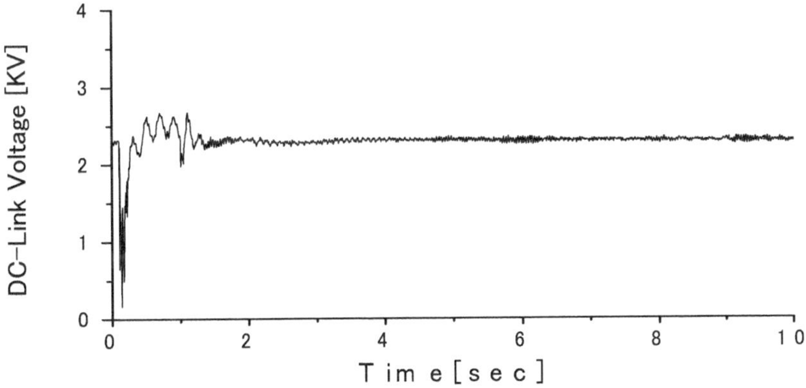

Fig. 8.19 DC-link circuit voltage (Case 2)

Case 3: In this case, an unsymmetrical 2LG fault is considered to occur at the sending end of one transmission line (fault point F2) of Fig. 8.6. The responses of the grid side reactive power, terminal voltage of the grid, real power, rotor speed of the PMSG, turbine blade pitch angle, and DC-link voltage are shown in Figs. 8.20 – 8.25, respectively. From the simulation results, it is clear that the proposed control system can also enhance the transient stability of a VSWT-PMSG under unsymmetrical fault condition.

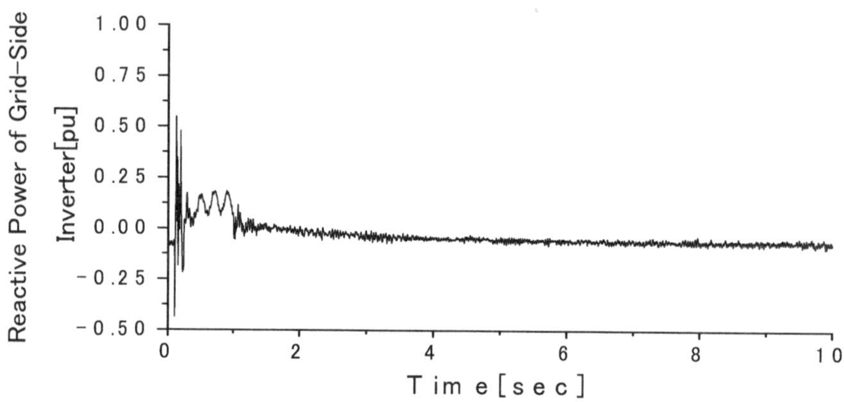

Fig. 8.20 Reactive power of the grid side inverter (Case 3)

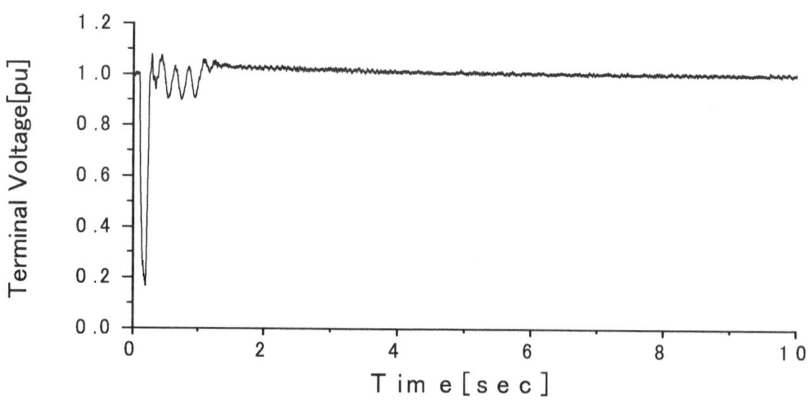

Fig. 8.21 Terminal voltage of the grid (Case 3)

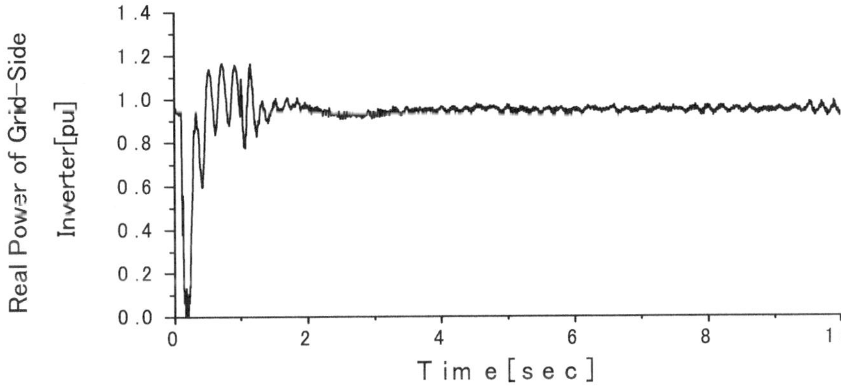

Fig. 8.22 Real power of the grid side inverter (Case 3)

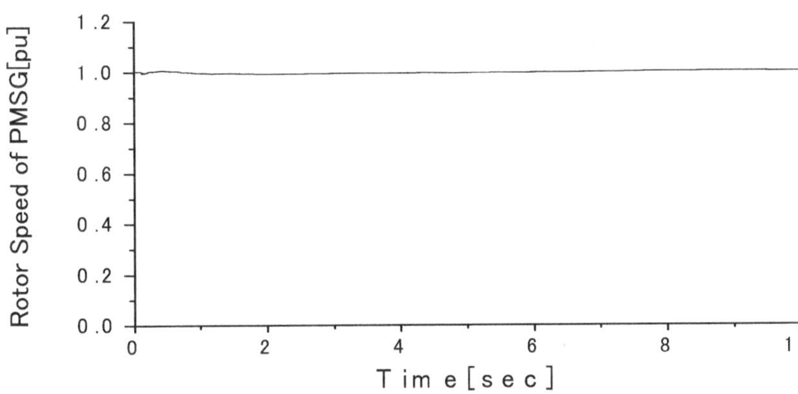

Fig. 8.23 Rotor speed of the PMSG (Case 3)

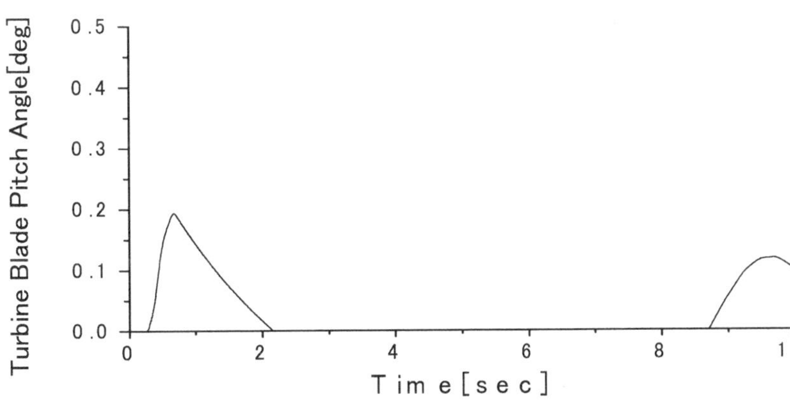

Fig. 8.24 Turbine blade pitch angle (Case 3)

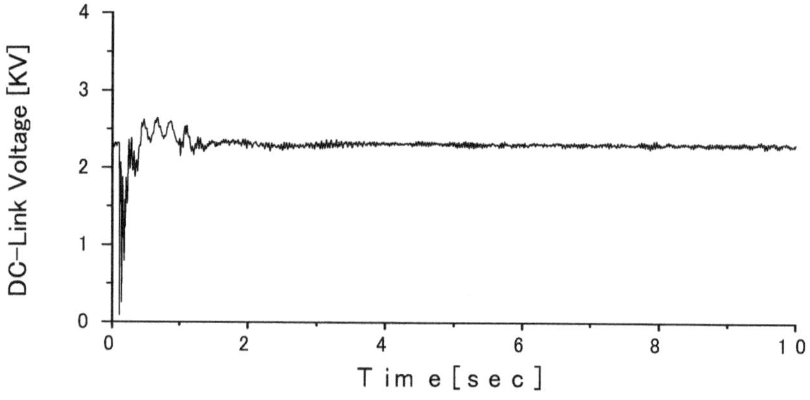

Fig. 8.25 DC-link circuit voltage (Case 3)

8.4 VSWT-PMSG with Rectifier-DC Chopper-DC Link-Inverter Topology

In this section, the direct drive VSWT-PMSG concept is analyzed with another type of fully controlled frequency converter composed of a generator side rectifier, DC chopper, DC-link, and grid side DC/AC inverter. The DC/AC inverter is a standard three-phase two-level unit, composed of six IGBTs and antiparallel diodes. The electrical scheme of the VSWT-PMSG topology is shown in Fig. 8.26. The control strategy of each converter is shown below.

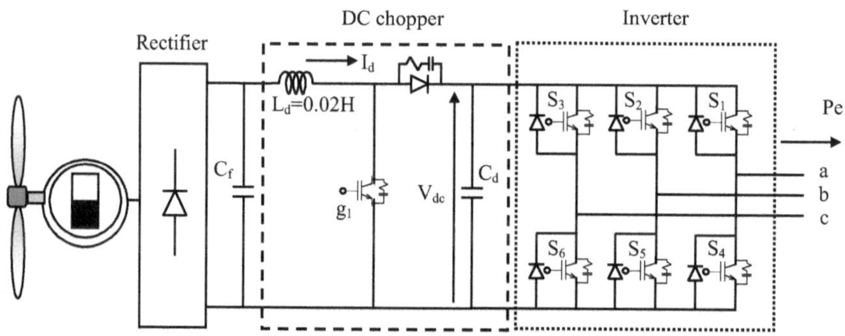

Fig. 8.26 Electrical scheme (2) of the VSWT-PMSG

8.4.1 Rectifier Topology

The AC output voltage of the PMSG is converted to the DC voltage by a diode rectifier circuit. C_f is a filter capacitance. Thyristor rectifiers and inverters require a constant current load on the DC side and an independent voltage source on the ac side because of the thyristor commutation process.

8.4.2 DC Chopper Control Strategy

The DC chopper is composed of an inductor, an IGBT switch, a diode, and the DC-link capacitor. Its purpose is to control the rectifier output current and thus the power. The gate signal is generated depending on the duty cycle, D, as shown in Fig. 8.27.

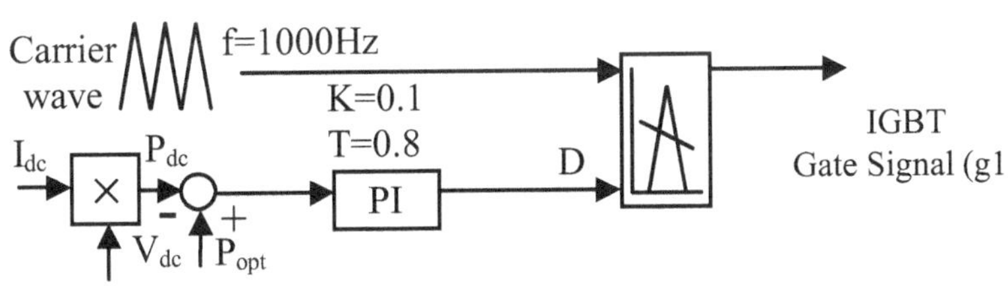

Fig. 8.27 Control block diagram of the DC chopper

8.4.3 Modeling and Control Strategy of Grid Side Inverter

The well-known cascaded control scheme is used for the grid side inverter. The grid side inverter modeling and control strategies are the same as those described in Sect. 8.2.2.

8.4.4 Model System Used in Sect. 8.4

The model system used for the transient stability analysis of the VSWT-PMSG is shown in Fig. 8.28, where a PMSG is connected to an infinite bus through a step-up transformer, and a double circuit transmission line. The PMSG field excitation may need to be strong when the rectifier-DC chopper-DC link-inverter topology is used, compared to the converter-DC link-inverter topology. In this work, the field flux is considered to be 1.55 pu. The other parameters of the PMSG used in the simulation are the same as those mentioned in Table 8.1. The system base is 5.0MVA.

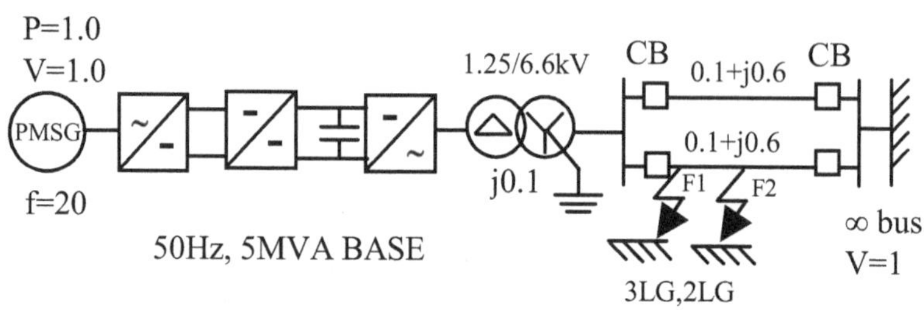

Fig. 8.28 Model system used in Sect. 6.4

8.4.5 Simulation Analysis

The time step and simulation time were chosen as 0.00001 sec and 10 sec, respectively. The fault occurs at 0.1 sec, the circuit breakers (CB) on the faulted lines are opened at 0.2 sec, and at 1.0 sec the circuit breakers are re-closed. In the transient stability analysis, the wind speed is kept constant at the rated speed at which the PMSG reference power is at the rated level, assuming that the wind speed doesn't change dramatically within this short time. The grid side converter parameters are shown in Table 8.4. Simulations were done by using PSCAD/EMTDC[3] [126]. For detailed LVRT and transient stability analysis of the VSWT-PMSG, three cases are considered as described below.

Table 8.4 PI controller parameters of the grid side inverter shown in Sect. 8.4.3

	PI-1	PI-2	PI-3	PI-4	PI-5
K_p	3.0	1.0	6.0	2.0	0.3
T_i	0.5	0.01	0.9	0.5	0.01

Case I: In this case, a 3LG fault is considered to occur at the middle of one transmission line (fault point F2) of Fig. 8.28. The grid side inverter can provide the necessary reactive power during the network disturbance as shown in Fig. 8.29. Therefore, the terminal voltage can return to its pre-fault level, as shown in Fig. 8.30. The response of the PMSG rotor speed is shown in Fig. 8.31. Depending

[3] For the latest information on PSCAD/EMTDC, visit at http://pscad.com

on the rotor speed the reference of the DC chopper is determined, as shown in Fig. 8.32. The real power response of grid is shown in Fig. 8.33. The pitch controller is activated when the rotor speed exceeds the nominal speed of the PMSG. The turbine blade pitch angle is shown in Fig. 8.34. The response of the DC-link voltage is shown in Fig. 8.35. From the simulation results, it is seen that the proposed control system can enhance the transient stability of the VSWT-PMSG when a 3LG fault occurs far from the wind generator.

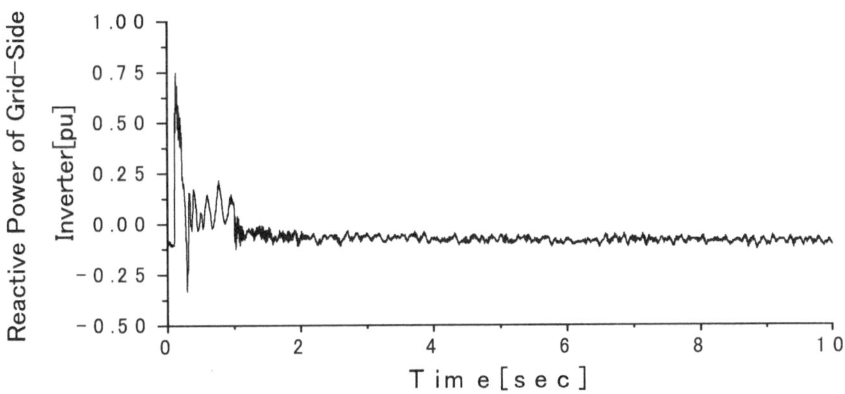

Fig. 8.29 Reactive power of the grid side inverter (Case I)

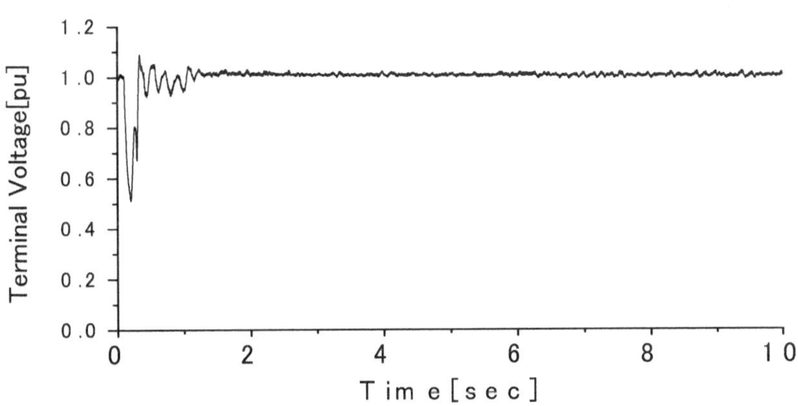

Fig. 8.30 Terminal voltage of the grid (Case I)

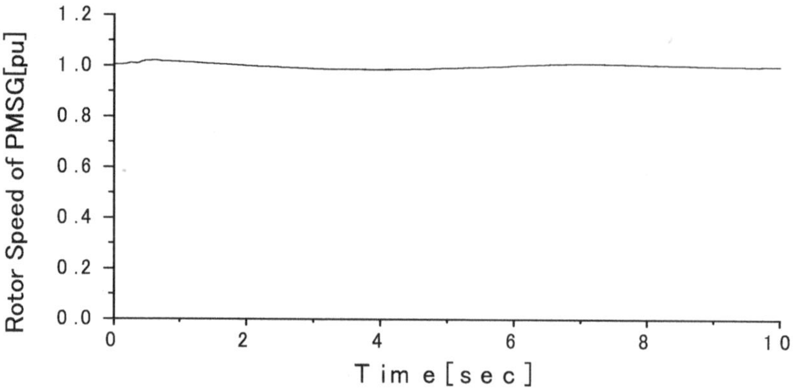

Fig. 8.31 Rotor speed of the PMSG (Case I)

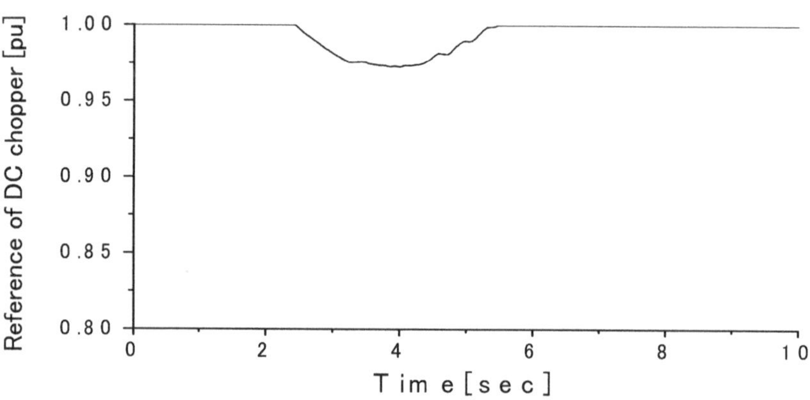

Fig. 8.32 Real power reference of the DC chopper (Case I)

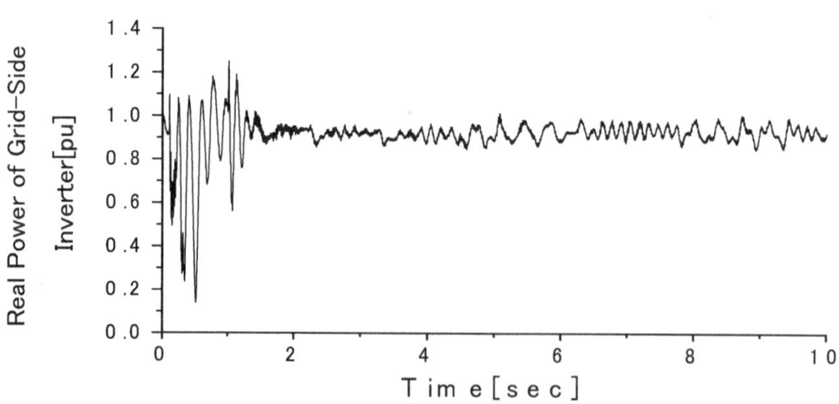

Fig. 8.33 Real power of grid side inverter (Case I)

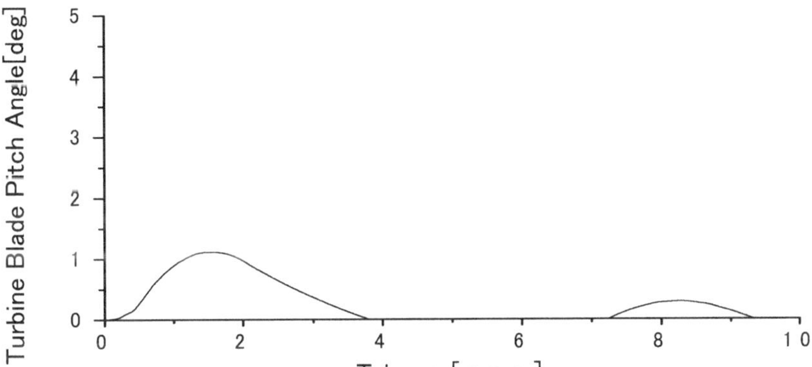

Fig. 8.34 Turbine blade pitch angle (Case I)

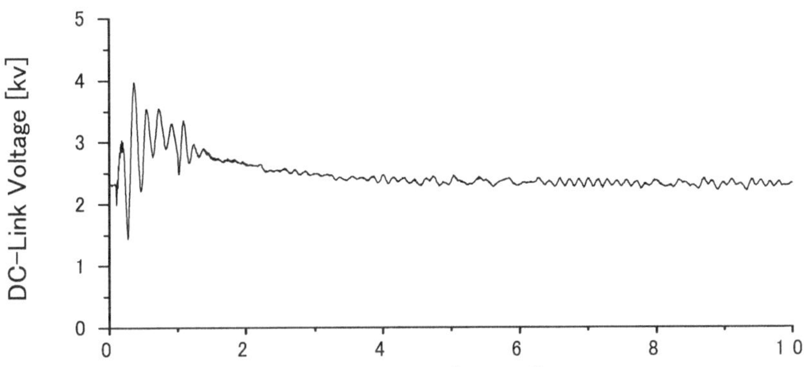

Fig. 8.35 DC-link circuit voltage (Case I)

Case II: In this case, a 3LG fault is considered to occur at the sending end of one transmission line (fault point F1) of Fig. 8.28. The responses of the grid side reactive power, terminal voltage of the grid, real power, rotor speed of the PMSG, turbine blade pitch angle, and DC-link voltage are shown in Figs. 8.36 – 8.41, respectively. From these results, it is seen that the proposed control system can also enhance the transient stability of a VSWT-PMSG when a 3LG fault occurs close to the wind generator.

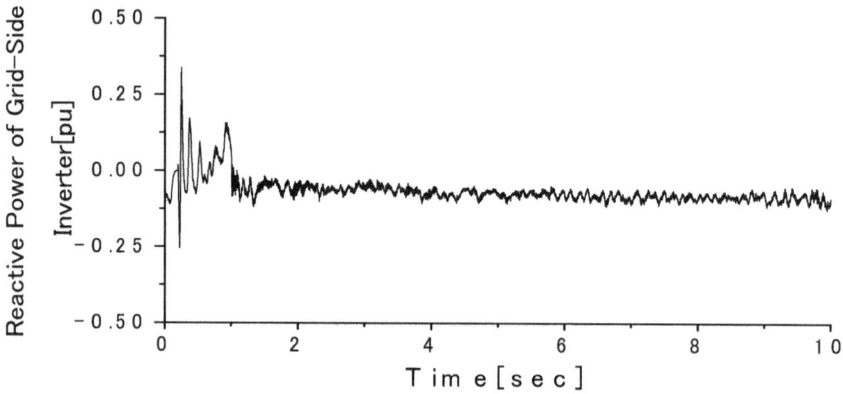

Fig. 8.36 Reactive power of grid side inverter (Case II)

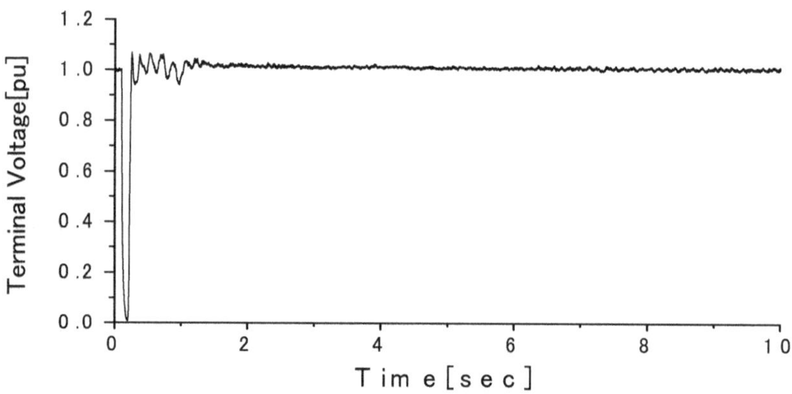

Fig. 8.37 Terminal voltage of the grid (Case II)

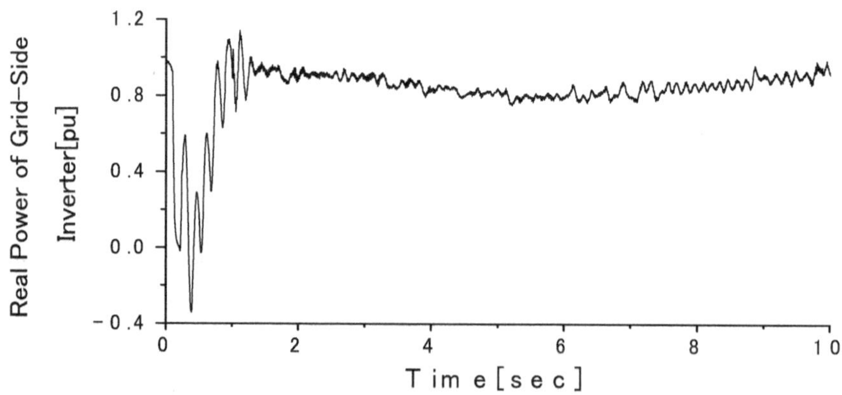

Fig. 8.38 Real power of grid side inverter (Case II)

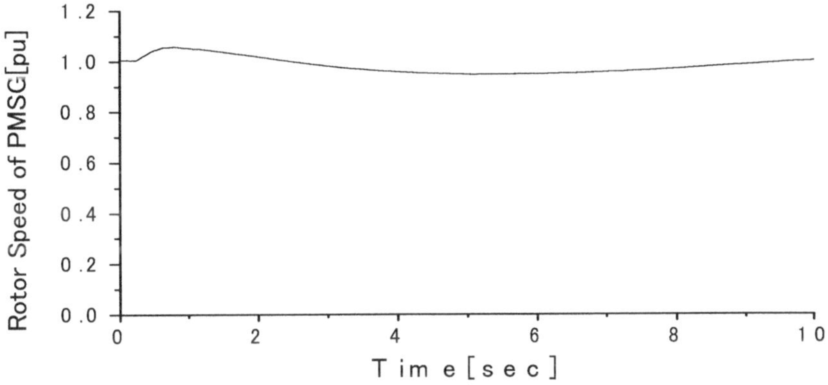

Fig. 8.39 Rotor speed of the PMSG (Case II)

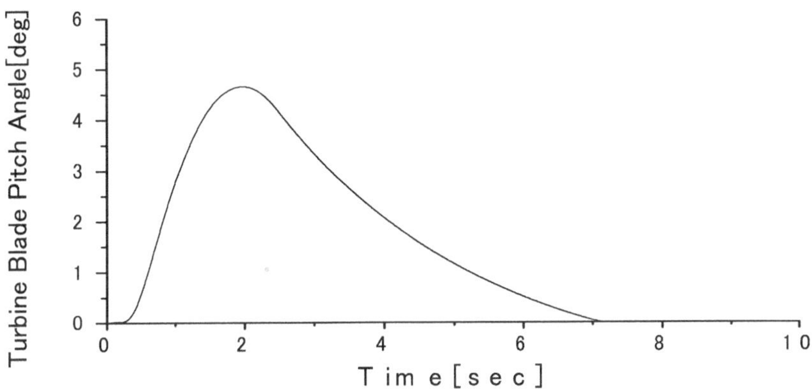

Fig. 8.40 Turbine blade pitch angle (Case II)

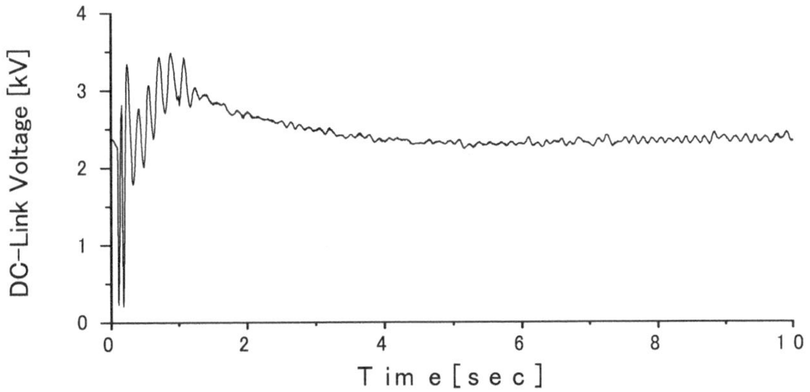

Fig. 8.41 DC-link circuit voltage (Case II)

Case III: In this case, an unsymmetrical 2LG fault is considered to occur at the sending end of one transmission line (fault point F1) of Fig. 8.28. The responses of the grid side reactive power, terminal voltage of the grid, real power, rotor speed of the PMSG, turbine blade pitch angle, and DC-link voltage are shown in Figs. 8.42 – 8.47, respectively. From the simulation results, it is clear that the proposed control system can also enhance the transient stability of a VSWT-PMSG under unsymmetrical fault condition.

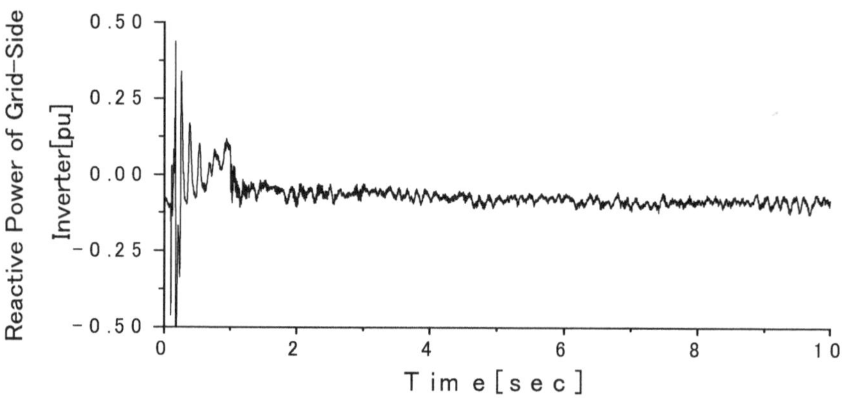

Fig. 8.42 Reactive power of grid side inverter (Case III)

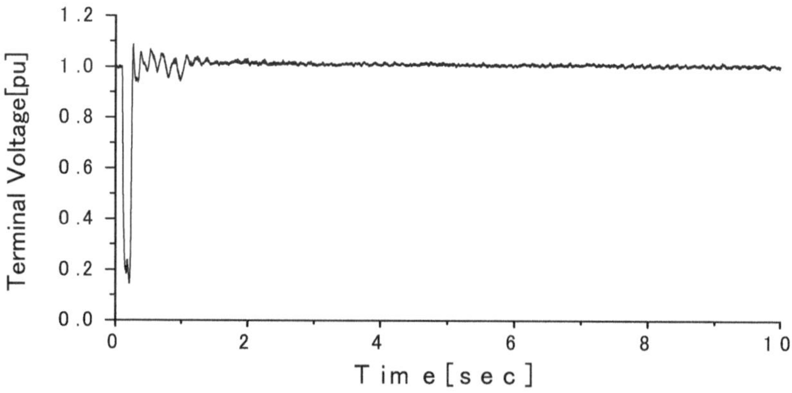

Fig. 8.43. Terminal voltage of the grid (Case III)

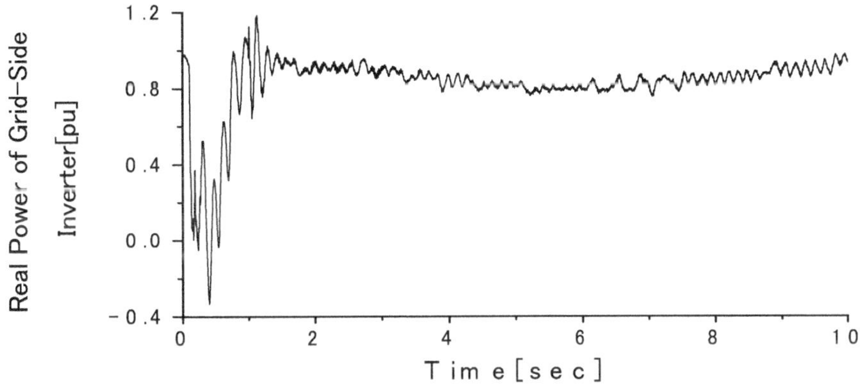

Fig. 8.44 Real power of the grid side inverter (Case III)

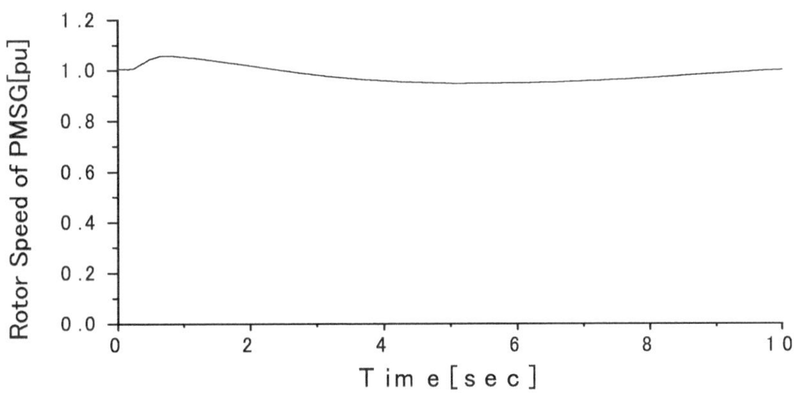

Fig. 8.45 Rotor speed of the PMSG (Case III)

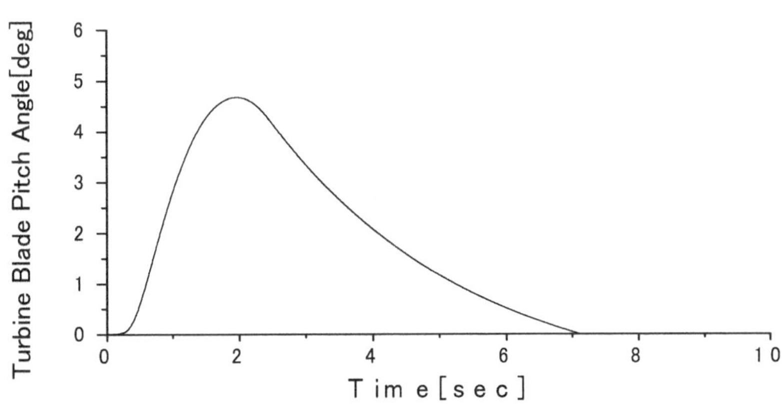

Fig. 8.46 Turbine blade pitch angle (Case III)

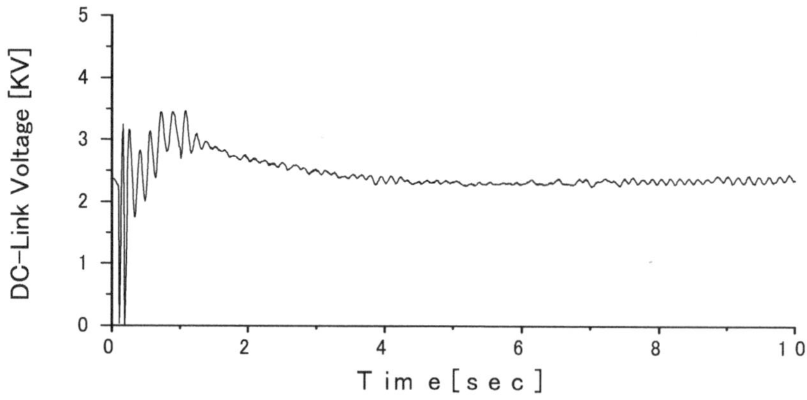

Fig. 8.47 DC-link circuit voltage (Case III)

8.5 Chapter Summery

This chapter presents a detailed study of the transient stability of the variable speed wind turbine driving a PMSG when a network disturbance occurs in the power system. First, detailed modeling of the wind turbine and the maximum power point tracking are described. Then two types of frequency converter topologies suitable for the VSWT-PMSG are presented. Then the modeling and control strategy for the generator and frequency converters are presented. The proposed control strategies can provide maximum power to the grid and can also control the reactive power to maintain the terminal voltage of the grid constant. These control strategies are suitable for improving the transient characteristics, where necessary reactive power is supplied, depending on the grid terminal voltage.

Finally, simulation results are shown using both types of topologies. Both symmetrical and unsymmetrical faults are considered as the network disturbances. It is found that a fault occurring near the generator side converter is more severe than a fault occurring far from the generator. Finally, it can be concluded that the proposed control system can increase the low voltage ride through (LVRT) capability of the VSWT-PMSG and thus the wind generator shutdown phenomenon during network disturbances can be decreased.

Acknowledgments Special thanks to Mr. Tomoki Asao for his great help in editing this chapter.

Appendix

A.1 Derivation of Eq. 4.1

The grid side voltage phasor, $\overline{V}_k$, is synchronized with the controller reference frame by using the phase locked loop (PLL). Therefore, if we look from the controller side, then the angle of the grid side voltage phasor seems to be zero. In that case, Eqs. A.1a – A.1e can be written.

$$\bar{I} = \frac{V_k - (V_{cd} + jV_{cq})}{R + jX} \tag{A.1a}$$

$$I_d = \frac{1}{R^2 + X^2}[R(V_k - V_{cd}) - XV_{cq}] \tag{A.1b}$$

$$I_q = \frac{-1}{R^2 + X^2}[X(V_k - V_{cd}) + RV_{cq}] \tag{A.1c}$$

$$P = \mathrm{Re}(\overline{V}_k \bar{I}^*) = V_k I_d \tag{A.1d}$$

$$Q = \mathrm{Im}(\overline{V}_k \bar{I}^*) = -V_k I_q \tag{A.1e}$$

From Eqs. A.1d and A.1e, Eq. A.2 can be obtained.

$$\begin{cases} P \propto I_d \\ Q \propto -I_q \end{cases} \tag{A.2}$$

If R<<X (as a winding resistance of transformer is much smaller than the leakage reactance), then from (A.1b) and (A.1c), we can get Eqs. A.3 and A.4:

$$\begin{cases} I_d \propto -V_{cq} \\ I_q \propto V_{cd} \end{cases} \tag{A.3}$$

Finally,

$$\begin{cases} P \propto I_d \propto -V_{cq} \\ Q \propto -I_q \propto -V_{cd} \end{cases} \tag{A.4}$$

A.2 Electrolyzer Characteristic

The electrolyzer characteristic is chosen from a technical report on a high-purity hydrogen and oxygen generator (HHOG) [141]. The voltage-current characteristic is shown below.

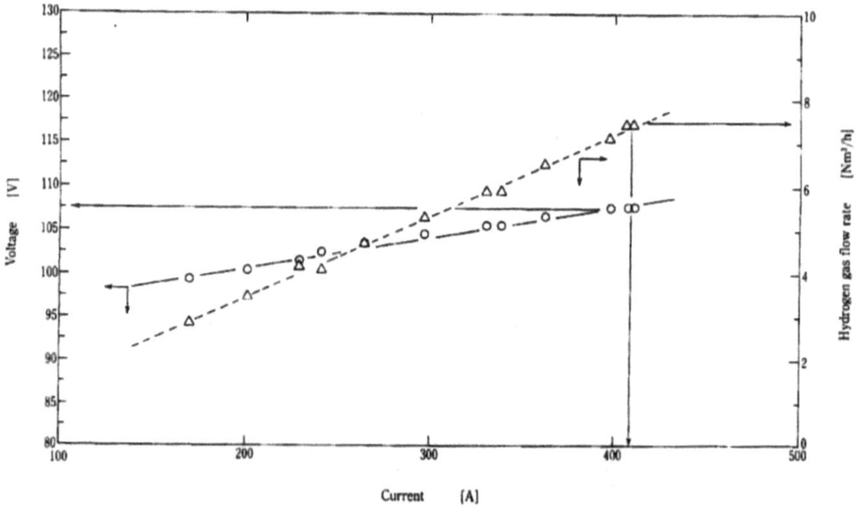

Fig. A.1 Relationship between voltage and current

From Fig. A.1, it is understood that the rated hydrogen gas flow rate is 7.5 Nm^3/h when the current and voltage values are 410 A and 107.5 V respectively.

Though there is a little non-linearity in the quantitative determination of hydrogen gas from an electric current, the linear approximation for the current-voltage characteristic doesn't give a large error. Therefore, in this study, we considered the linear approximation for the current-voltage characteristic to simulate the hydrogen electrolyzer.

Now, two points {(410 A, 107.5 V) and (300 A, 104.1 V)} are chosen from Fig. A.1.

Then the following linear function (Eq. A.5) can be developed easily:

$$\begin{aligned} V &= 0.031 \times I + 94.8 \\ \Rightarrow V &= R_0 \times I + V_0 \end{aligned} \quad \text{(A.5)}$$

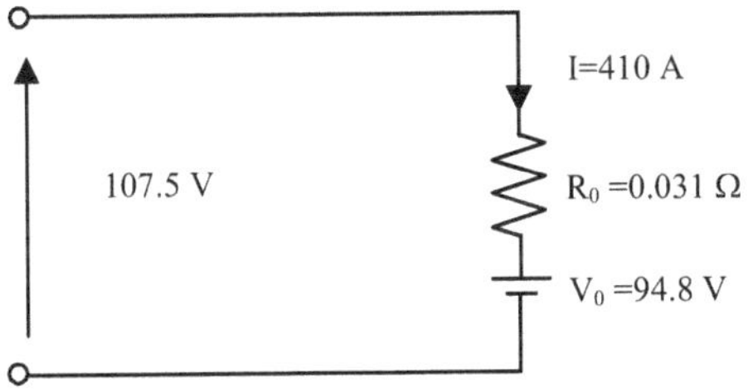

Fig. A.2 Equivalent circuit of an electrolyzer cell

From Eq. A.5, one electrolyzer cell can be expressed easily as shown in Fig. A.2, which is composed of an electromotive force, V_0 and an internal resistance, R_0. During the rated operation, the electrolyzer consumes the rated power of 44.075 kW.

References

1. GWEC Latest News (2008) US, China & Spain lead world wind power market in 2007. The Global Wind Energy Council, Feb.
http://www.gwec.net/. Cited 28 Feb 2008.
2. AWEA Resources (2008) U.S. Wind Energy Projects. The American Wind Energy Association, Jan.
http://www.awea.org/. Cited 10 Feb 2008.
3. EWEA Publications (2008) Wind energy leads EU power installations in 2007, but national growth is inconsistent. The European Wind Energy Association, Feb.
http://www.ewea.org/. Cited 27 Feb 2008.
4. EWEA Publications (2007) Special edition of Wind Directions: EWEA's anniversary – The first 25 years, Sep/Oct, 2007. The European Wind Energy Association.
http://www.ewea.org/. Cited 25 Feb 2008.
5. Enercon Homepage.
http://www.enercon.de/en/_home.htm. Cited 27 Feb 2008.
6. F. Van Hulle (2005) Large Scale Integration of Wind Energy in the European Power Supply: Analysis, Issues, and Recommendations. The European Wind Energy Association (EWEA) report.
http://www.ewea.org/. Cited 20 Feb 2008.
7. R. Zavadil, N. Miller, A. Ellis, E. Muljadi (2005) Making connections. In: IEEE Power & Energy Magazine, Vol.3, No.6, pp.30 – 32.
8. Federal Energy Regulatory Commission (FERC) report (2005) Interconnection for wind energy. FERC, United States of America, Docket No. RM05-4-000 – Order No. 661, 2 Jun.
9. Ireland Grid Code Version 2 (2007), Wind farm power station grid code provisions, WFPS1. From: Ireland National Grid report, pp.213 – 216, Jan.
10. C.L. Souza et al. (2001) Power system transient stability analysis including synchronous and induction generator. In: IEEE Porto Power Tech Proceedings, Vol.2, pp.6.
11. E.S.Abdin, W. Xu (2000) Control design and dynamic performance analysis of a wind turbine induction generator unit. In: IEEE Trans. on Energy Conversion, Vol.EC – 15, No.1, p.91.
12. I. Zubia et al. (2001) Electrical fault simulation and dynamic response of a wind farm. In: Proc. of thc IASTED International Conference Power and Energy System, No.337 – 095, pp.595 – 600.
13. J. Tamura et al. (2001) Transient stability simulation of power system including wind generator by PSCAD/EMTDC. In: IEEE Porto Power Tech Proceedings 2001, Vol.4, EMT – 108.
14. J. Tamura et al. (2002) Analysis of transient stability of wind generators. In: Conference Record of ICEM 2002, No.148.

15. E.N. Hinrichsen et al. (1982) Dynamics and stability of wind turbine generators. In: IEEE Trans. on Power Apparatus and System, Vol. PAS - 101, No.8, p.2640.
16. J. Tamura, Y. Shima, R. Takahashi, T. Murata, Y. Tomaki, S. Tominaga, et al. (2005) Transient stability analysis of wind generator during short circuit faults. In: Proc. of 3rd IEEE International Conference on Systems, Signals & Devices (SSD'05), Keynote Lecture PES – 1, Sousse, Tunisia.
17. A. G. Gonzalez Rodriguez, M. Burgos Payan, C. Izquierdo Mitchell (2001) PSCAD based simulation of the connection of a wind generator to the network. In: IEEE Porto Power Tech Proceedings 2001, DRS3 – 307.
18. C. Carrillo et al. (2004) Power fluctuations in an isolated wind plant. In: IEEE Trans. On Energy Conversion, Vol.19, No.1, p.217.
19. T. Petru, T. Thiringer (2002) Modeling of wind turbines for power system studies. In: IEEE Trans. on Power Systems, Vol.7, No.4, pp.1132 – 1139.
20. V. Akhmatov et al. (2003) Modelling and transient stability of large wind farms. In: Int. J. Elect. Power and Energy Syst., Vol.25, pp.123 – 144.
21. V. Akhmatov, H. Knudsen, A. H. Nielsen (2000) Advanced simulation of windmills in the electric power supply. In: Int. J. Elect. Power and Energy Syst., Vol.22, pp.421 – 434.
22. Y. Shima, et al. (2005) Transient stability simulation of wind generator expressed by two – mass model. In: IEEJ Trans. on PE., Vol.125 – B, No.9.
23. S.M. Muyeen, M.A. Mannan, M.H. Ali, R. Takahashi, T. Murata, J. Tamura, et al. (2005) Transient stability analysis of wind generator system with the consideration of multi – mass shaft model. In: International Conference on Power Electronics and Drive Systems (IEEE PEDS 2005), Conference CDROM, pp.511 – 516, Malaysis.
24. S.K. Salman et al. (2001) Improvement of fault clearing time of wind farm using reactive power compensation. In: IEEE Porto Power Tech Proceedings 2001, SSR2 – 097.
25. S.K. Salman et al. (2002) Investigation into the estimation of the critical clearing time of a grid connected wind power based embedded generator. In: Proc. IEEE/PES T & D Conference and Exhibition 2002 Asia pacific, Vol.II, p.975.
26. S.K. Salman, Anita L.J. Teo (2003) Windmill modeling consideration and factors influencing the stability of a grid – connected wind power – based embedded generator. In: IEEE Trans. on Power Systems, Vol.18, No.2, pp.793 – 802.
27. D.J. Trudnowski et al. (2004) Fixed – speed wind – generator and wind – park modeling for transient stability studies. In: IEEE Trans. on Power Systems, Vol.19, No.4, p.1911 – 1917.
28. W.E. Leithead, B.Connor (2000) Control of variable speed wind turbines: dynamic models. In: Int. J. Control, Vol.73, No. 13, pp.1173 – 1188.
29. L. Mihet – Popa et al. (2004) Wind turbine generator modeling and simulation where rotational speed is the controlled variable. In: IEEE Trans. on Industry Applications, Vol.40, No.1, pp.3 – 10.
30. T. Thiringer et al. (1996) Power quality measurements performed on a low – voltage grid equipped with two wind turbines. In: IEEE Trans. on Energy Conversion, Vol.11, No.3, pp.601 – 606.
31. S.A. Papathanassiou et al. (2001) Mechanical stresses in fixed speed wind turbines due to network disturbances. In: IEEE Trans. on Energy Conversion, Vol.16, No.4, pp.361 – 367.
32. S.A. Papathanassiou (1997) Contribution to the analysis of variable speed wind turbines with induction generator for the selection of their electrical scheme. PhD Thesis, National Technical University of Athens (NTUIA).
33. O. Wasynczuk, D.T. Man, J.P. Sullivan (1981) Dynamic behavior of a class of wind turbine generator during random wind fluctuations. In: IEEE Trans. on Power Apparatus and Systems, Vol. PAS-100, No.6, pp.2873 – 2845.
34. J.G. Slootweg, S.W.H. De Haan (2003) General model for representing variable speed wind turbines in power system dynamics simulation. In: IEEE Trans. on Power Systems, Vol.18, Issue 1, pp.144 – 151.

35. A. Murdoch, J.R. Winkelman, S. H. Javid (1983) Control design and performance analysis of a 6 MW wind turbine - generator. In: IEEE Trans. on Power Apparatus and System, Vol.PAS-102, No.5, pp.1340 – 1347.
36. E.N. hinrichsen (1983) Controls for variable pitch wind turbine generators. In: IEEE Trans. on Power Apparatus and Systems, Vol. PAS-103, No.4, pp. 886 – 892.
37. E. Muljadi, C.P. Butterfield (2001) Pitch - controlled variable - speed wind turbine generation. In: IEEE Trans. on Industry Applications, Vol.37, No.1, pp.240 – 246.
38. M.H. Hansen, A. Hansen, T.J. Larsen, S. Oye, P. Sorensen, P. Fuglsang (2005) Control design for a pitch - regulated variable speed wind turbine. Riso National Laboratory, Riso - R – 1500, Denmark.
39. F.J.L. Van Hulle, C. Nath, P.H. Jensen, C. Eriksson, P. Vionis (2001) European wind turbine certification EWTC - guidelines for design evaluation of wind turbines. Energy Research Center of the Netherlands (ECN) report, ECN-C--1-059.
40. T. Senjyu, R. Sakamoto, T. Kinjo, K. Uezato, T. Funabashi (2004) Output power leveling of wind turbine generator by pitch angle control using generalized predictive control. In: The Papers of Joint Technical Meeting on Power Engineering and Power Systems Engineering, IEE Japan, PE-04-77/PSE-04-77, pp.17 – 22, in Japanese.
41. O. Kanna, S. Hanba, S. Asato, K. Yamashita (1997) A method of stabilization of a wind generator power using backstepping algorithm. In: IEEJ Trans. on PE., Vol.117-B, No.12, pp.1513 – 1519, in Japanese.
42. M.M. Hand and M.J. Balas (2002) Systematic controller design methodology for variable-speed wind turbines. National Renewable Energy Laboratory (NREL) report, NREL/TP-500-29415, Golden, Colorado, USA.
43. S.M. Muyeen, R. Takahashi, T. Murata, J. Tamura (2005) Transient stability enhancement of wind generator by online logical controller with the consideration of initial condition settings. In: The Proceedings of the IPEC (International Power Electronics Conference) Nigata, Japan.
44. H.F. Wang, F. Li, R.G. Cameron (1999) Facts control design based on power system nonparametric models. In: IEE Proc.-Gener. Transm. Distrib., Vol.146, No.5, pp.409 – 415.
45. L. Gyugyi (1994) Dynamic compensation of AC transmission lines by solid-state synchronous voltage sources. In: IEEE Trans. on Power Delivery, Vol.9, No.2, pp.904 – 911.
46. L. Gyugyi (1992) Unified power-flow control concept for flexible AC transmission systems. In: IEE Proc.-C, Vol.139, No.4, pp.323 – 331.
47. L. Cong, Y. Wang, D.J. Hill (2005) Transient stability and voltage regulation enhancement via coordinated control of generator excitation and SVC. In: Electrical Power and Energy Systems, Vol.27, pp.121 – 130.
48. T. Ahmed, O. Noro, E. Hiraki, M. Nakaoka (2004) Terminal voltage regulation characteristics by static var compensator for a three-phase self-excited induction generator. In: IEEE Trans. on Industry Applications, Vol.40, No.4, pp.978 – 988.
49. Y.L. Tan (1999) Analysis of line compensation by shunt-connected FACTS controllers: A comparison between SVC and STATCOM. In: IEEE Power Engineering Review, pp.57-58.
50. K.R. Padiyar, A.M. Kulkarni (1997) Design of reactive current and voltage controller of static condenser. In: Electric Power & Energy Systems, Vol.19, No.6, pp.397 – 410.
51. C.K. Sao, P.W. Lehn, M.R. Iravani, J. A. Martinez (2002) A benchmark system for digital time-domain simulation of a pulse-width-modulated D-STATCOM. In: IEEE Trans. on Power Delivery, Vol.17, No.4, pp.1113 – 1120.
52. G.W. Moon (1999) Predictive current control of distribution static compensator for reactive power compensation. In: IEE Proc. Gener. Transm. Distrib., Vol.146, No.5, pp.515 – 520.
53. H.F. Wang (1999) Phillips-hoffron model of power systems installed with STATCOM and applications. In: IEE Proc.-Gener. Transm. Distrib., Vol.146, No.5, pp.521 – 527.

54. C. Schauder, M. Gernhardt, E. Stacey, T. Lemak, L. Gyugyi, T.W. Cease, A. Edris (1995) Development of a ± 100MVAR static condenser for voltage control of transmission systems. In: IEEE Trans. on Power Delivery, Vol.10, No.3, pp.1486 – 1496.
55. O. Anaya-Lara, E. Acha (2002) Modeling and analysis of custom power system by PSCAD/EMTDC. In: IEEE Trans. on Power Delivery, Vol.17, No.1, pp.266 – 272.
56. Z. Saad-Saoud (1998) Application of STATCOMs to Wind Farms. In: IEE Proc. Gener. Transm. Distrib., Vol.145, No.5, pp.511 – 517.
57. Z. Chen, F. Blaabjerg, Y. Hu (2005) Voltage recovery of dynamic slip control wind turbines with a STATCOM. In: International Power Electronic Conference (IPEC05), S29-5, pp.1093 – 1100.
58. T. Sun, Z. Chen, F. Blaabjerg (2004) Flicker mitigation of grid connected wind turbines using STATCOM. In: Proc. of 2nd IEE International Conference on Power Electronics, Machines and Drives, PEMD04.
59. T. Larsson, C. Poumarede (1999) STATCOM, an efficient means for flicker mitigation. In: Proceedings of IEEE Power Engineering Society 1999 Winter Meeting, pp.1208 – 1213, New York, USA.
60. S.M. Muyeen, M.A. Mannan, M.H. Ali, R. Takahashi, T. Murata, J. Tamura (2005) Stabilization of grid connected wind generator by STATCOM. In: International Conference on Power Electronics and Drive Systems (IEEE PEDS 2005), Conference CDROM, pp.1584 – 1589, Malaysia.
61. G.C. Cho, G.H. Jung, N.S. Choi, G.H. Cho (1996) Analysis and controller design of static var compensator using three-level GTO inverter. In: IEEE Trans. On Power Electronics, Vol.11, No.1, pp.57 – 65.
62. M.I. Mazurov, A.V. Nikolaev, N.G. Lozinova (2005) Improvement of voltage quality in AC network by use of STATCOM. In: IEEE PowerTech 2005, Russia.
63. J.B. Ekanayake, N. Jenkins (1997) Mathemetical models of a three-level advanced static VAR compensator. In: IEE Proc.-Gener. Transm. Distrib., Vol.144, No.2, pp.201 – 206.
64. A. Abolins, D. Lambrecht, J. S. Joyce, L. T. Rosenberg (1976) Effect of clearing short circuit and automatic reclosing on torsional stress and life expenditure of turbine-generator shafts. In: IEEE Trans. on Power Apparatus and Systems, Vol. PAS-95, No.1.
65. O. Wasynczuk (1981) Damping shaft torsional oscillations using a dynamically controlled resistor bank. In: IEEE Trans. on Power Apparatus and Systems, Vol. PAS-100, No.7.
66. R. Cardenas, R. Pena, G. Asher, J. Clare (2004) Power smoothing in wind generation systems using a sensorless vector controlled induction machine driving a flywheel. In: IEEE Trans. on Energy Conversion, Vol.19, Issue.1, pp. 206 – 216.
67. F. Hardan, J.A.M Bleijs, R. Jones, P. Bromley, A.J. Ruddell (1999) Application of a power-controlled flywheel drive for wind power conditioning in a wind/diesel power system. In: 9th International Conference on Electrical Machines and Drives, paper No. 468, pp.65 – 70, Canterbury.
68. R. Takahashi, Wu. Li, T. Murata, J. Tamura (2005) An application of flywheel energy storage system for wind energy conversion. In: International Conference on Power Electronics and Drives Systems, (PEDS 2005), pp. 932 – 937.
69. T. Matsuzaka, K. Tuchiya (1997) Study on stabilization of a wind generator power fluctuation. In: IEEJ Trans. on PE., Vol.117-B, No.5, pp.625 – 633,in Japanese.
70. T. Senjyu, R. Sakamoto, N. Urasaki, H.Higa, K. Uezato, T. Funabashi (2004) Output power leveling of wind turbine generator by pitch angle control using minimum variance control. In: IEEJ Trans. on PE., Vol.124-B, No.12, pp.1455 – 1462, in Japanese.
71. T. Senjyu, R. Sakamoto, N. Urasaki, T. Funabashi, H. Fujita, H. Sekine (2005) Output power leveling of wind turbine generators using pitch angle control for all operating regions in wind farm. In: IEEJ Trans. on PE., Vol.125-B, No.12, pp.1159 – 1168, in Japanese.
72. T. Senjyu, R. Sakamoto, N. Urasaki, T. Funabashi, H. Fujita, H. Sekine (2006) Output power leveling of wind turbine Generator for all operating regions by pitch angle control. In: IEEE Trans. on Energy Conversion, Vol.21, Issue.2, pp.467 – 475.

73. F. Zhou, G. Joos, C. Abbey, L. Jiao, B.T. Ooi (2004) Use of large capacity SMES to improve the power quality and stability of wind farms. In: IEEE Power Engineering Society General Meeting, Vol.2, pp. 2025 – 2030.
74. S. Nomura, Y. Ohata, T. Hagita, H. Tsutsui, S. Tsuji-Lio, R. Shimada (2005) Wind farms linked by SMES systems. In: IEEE Trans. on Applied Superconductivity, Vol. 15, Issue.2, part2, pp. 1951 – 1954.
75. M.H. Ali, T. Murata, J. Tamura (2006) Minimization of fluctuations of line power and terminal voltage of wind generator by fuzzy logic-controlled SMES. In: International Review of Electrical Engineering (I.R.E.E.), Vol.1, No.4, pp.559 – 566.
76. L. Zhang, C. Shen, M.L. Crow, L. Dong, S. Pekarek, S. Atcitty (2005) Performance indices for the dynamic performance of FACTS and FACTS with energy storage. In: Electric Power Component and System, Vol.33, No.3, pp.299 – 314.
77. Z. Yang, C. Shen, L. Zhang, M.L. Crow, S. Atcitty (2001) Integration of a STATCOM and battery energy storage. In: IEEE Trans. on Power Syst., Vol.16, No.2,pp.254 – 260.
78. A. Arulampalam, J.B. Ekanayake, N. Jenkins (2003) Application study of a STATCOM with energy storage. In: Proc. Inst. Electr. Eng.-Gener. Transm. Distrib., vol.150,no.3,pp.373 – 384.
79. Y. Cheng, C. Qian, M. L. Crow, S. Pekarek, S. Atcitty (2006) A comparison of diode-clamped and cascaded multilevel converters for a STATCOM with energy storage. In: IEEE Trans. on Industrial Electronics, Vol. 53, No. 5, pp.1512 – 1521.
80. S.M. Muyeen, M.H. Ali, R. Takahashi, T. Murata, J. Tamura (2007) Wind generator output power smoothing and terminal voltage regulation by using STATCOM/ESS. In: CD Record of the IEEE PowerTech 2007 conference, Ref. No. 258, Lausanne, Switzerland.
81. E. Spahic, G. Balzer, B. Hellmich, W. Munch (2007) Wind energy storage-possibilities. In: CD Record of the IEEE PowerTech 2007 conference, Paper No. 123, Lausanne, Switzerland.
82. M. Okamura (1995) A basic study on Power Storage Capacitor System. In: IEEJ Trans. on PE., Vol.115-B, pp.504 – 510, in Japanese.
83. L.S. Russell, R.M. Nelms (2000) Optimization of double-layer capacitor arrays. In: IEEE Trans. on Industry Applications, Vol.36, No.1, pp.194 – 198.
84. Y. Nozaki, K. Akiyama, H. Kawaguchi, K. Kurokawa (2000) An improved method for controlling an EDLC-battery hybrid stand-alone photovoltaic power system. In: Applied Power Electronics Conference and Exposition (APEC 2000), Vol.2, pp.781 – 786.
85. Y. Jia, R. Shibata, N. Yamamura, M. Ishida (2005) A control method of prolonging the service life of battery in stand-alone renewable energy system using electric double layer capacitor (EDLC). In: IEEE conference on Power Electronics and Drives Systems, (PEDS 2005). Vol.1, pp.228 – 233.
86. S. Sugimoto, I. Kouda, Y. Murai (1998) Energy storage system utilizing large capacity electric double-layer capacitors for peak-cut of power demand. In: IEEJ Trans., Vol.118-D, pp.1377 – 1385, in Japanese.
87. T. Muto (2000) Development technology of the instantaneous voltage sag compensator applied to a large-capacity electric double-layer capacitor. In: Electron Technology, Vol.44, No.11, pp.52 – 58, in Japanese.
88. G. Ariyoshi, K. Murata, K. Harada, K. Yamasaki (2000) Load levelling using EDLCs under PLL control. In: IEICE Trans. Fundam., Vol. E83-A, No.6, pp. 1014 – 1022.
89. L. Zubieta, R. Bonert (2000) Characterization of double-layer capacitors for power electronics applications. In: IEEE Trans. on Industry Applications, Vol. 36, No.1, pp.199-205.
90. T. Kinjo, T. Senjyu, N. Urasaki, H. Fujita (2006) Output leveling of renewable energy by electric double-layer capacitor applied for energy storage system. In: IEEE Trans. on Energy Conversion, Vol. 21, Issue.1, pp. 221 – 227.
91. L.J. Fingersh (2003) Optimized hydrogen and electricity generation from wind. National Renewable Energy Laboratory (NREL) report, NREL/TP-500-34364, pp.1 – 12.

92. R. Kottenstette, J. Cotrell (2003) Hydrogen storage in wind turbine towers. National Renewable Energy Laboratory (NREL) report, NREL/TP-500-34656, pp.1 – 21.
93. R. Dufo-Lopez, J.L. Bernal-Agustin, J. Contreras (2007) Optimization of control strategies for stand-alone renewable energy systems with hydrogen storage. In: Renewable Energy, Vol.32, No.7, pp.1102 – 1126.
94. H. Nakabayashi, R. Takahashi, T. Murata, J. Tamura, M. Futami, M. Ichinose, et al. (2006) Fundamental research of hydrogen production system using variable speed wind generator. In: XVII International Conference on Electrical Machines (ICEMS06), Conference CDROM Ref. No. 316, Greece.
95. P. Ledesma, J. Usaola (2005) Doubly fed induction generator model for transient stability analysis. In: IEEE Trans. on Energy Conversion, Vol.20, No.2, pp.388 – 397.
96. P.La Seta, P. Schegner (2005) Comparison of stabilizing methods for doubly-fed induction generators for wind turbines. In: International Conference on Future Power System, Conference CDROM.
97. T. Sun, Z. Chen, F. Blaabjerg (2005) Transient stability of DFIG wind turbines at an external short-circuit fault. In: Wind Energy, Vol.8, No.3, pp.345 – 360.
98. M.S. Morsy, H.H. Amer, M.A. Badr, A.M. El-Serafi (1983) Transient stability of synchronous generators with two-axis slip frequency excitation. In: IEEE Trans. on Power Apparatus and Systems, Vol.102, No.4, pp.852 – 859.
99. L. Feng, M. Shengwei, X. Deming, M. Yongjian, J. Xiaohua, L. Qiang (2004) Experimental evaluation of nonlinear robust control for SMES to improve the transient stability of power systems. In: IEEE Trans. on Energy Conversion, Vol.19, No.4, pp.774 – 782.
100. A. Yazdani, R. Iravani (2006) A neutral-point clamped converter system for direct-drive variable-speed wind power unit. In: IEEE Trans. on Energy Conversion, Vol.21, No.2, pp.596 – 607.
101. Peter Vas (1992) Electrical Machines and Drives- A Space Vector Theory Approach. Oxford University Press, New York, United States.
102. T.J.E. Miller (1989) Brushless Permanent-Magnet and Reluctance Motor Drives. Oxford University Press, New York, United States.
103. M. Chinchilla, S. Arnaltes, J. C. Busgos (2006) Control of permanent-magnet synchronous generators applied to variable-speed wind-energy systems connected to the grid. In: IEEE Trans. on Energy Conversion, Vol.21, No.1.
104. N.A. Cutululis, E. Ceanga, A.D. Hansen, P. Sørensen (2006) Robust multi-model control of an autonomous wind power system. In: Wind Energy, Vol.9, No.5, pp. 399 – 419.
105. K. Tan, S. Islam (2004) Optimum control strategies in energy conversion of PMSG wind turbine system without mechanical sensors. In: IEEE Trans. on Energy Conversion, Vol.19, No.2, pp.392 – 399.
106. S. Morimoto, T. Nakamura, Y. Takeda (2003) Power maximization control of variable-speed wind generation system using permanent magnet synchronous generator. In: IEEJ Trans. on P.E., Vol. 123-B, No.12, pp.1573 – 1579.
107. S.M. Muyeen, M.H. Ali, R. Takahashi, T. Murata, J. Tamura, Y. Tomaki, et al. (2006) Transient stability analysis of grid connected wind turbine generator system considering multi-mass shaft modeling. In: Electric Power Components & Systems, Vol.34, No.10, pp.1121 – 1138.
108. S.M. Muyeen, M.H. Ali, R. Takahashi, T. Murata, J. Tamura (2006) Transient stability enhancement of wind generator by a new logical pitch controller. In: IEEJ Trans.PE., Vol.126-B, No.08, pp.742 – 752.
109. S.M. Muyeen, M.A. Mannan, M.H. Ali, R. Takahashi, T. Murata, J. Tamura (2006) Stabilization of wind turbine generator system by STATCOM. In: IEEJ Trans.PE., Vol.126-B, No.10, pp.1073 – 1082.
110. S.M. Muyeen, M.H. Ali, R. Takahashi, T. Murata, J. Tamura, Y. Tomaki, et al. (2007) Analysis of wind generator transient stability considering six-mass drive train model. In: International Review of Electrical Engineering (I.R.E.E.), Vol.2, No.1, pp.91 – 102.

111. S.M. Muyeen, M.H. Ali, R. Takahashi, T. Murata, J. Tamura, Y. Tomaki, et al. (2007) A comparative study on transient stability analysis of wind turbine generator system using different drive train models. In: IET-Proceedings on Renewable Power Generation (IET-RPG), Vol.1, No.2, pp.131 – 141.
112. S.M. Muyeen, M.H. Ali, R. Takahashi, T. Murata, J. Tamura (2008) Damping of blade-shaft torsional oscillation of wind turbine generator system. In: Electric Power Components & Systems, Vol.36, No.2, pp. 195 – 211.
113. S.M. Muyeen, M.H. Ali, R. Takahashi, T. Murata, J. Tamura (2007) Wind generator output power smoothing by using pitch controller. In: International Review of Electrical Engineering (I.R.E.E.), Vol.2, No.3, pp.310 – 321.
114. S.M. Muyeen, S. Shishido, M.H. Ali, R. Takahashi, T. Murata, J. Tamura (2008) Application of energy capacitor system (ECS) to wind power generation. In: Wind Energy, Vol. 11, No. 4, pp. 335 – 350, DOI: 10.1002/we.265.
115. S.M. Muyeen, M.H. Ali, R. Takahashi, T. Murata, J. Tamura, M. Kubo, et al. (2008), "Low voltage ride through capability enhancement of wind turbine generator system during network disturbance," Accepted to publish in the IET Proceedings on Renewable Power Generation, Paper No. RPG-2007 – 0116.
116. G.L. Johnson, (2008) Wind Energy Systems. http://eece.ksu.edu/~gjohnson/. Cited 28 Feb 2008.
117. E. Golding (1976) The generation of electricity by wind power. Halsted Press, New York.
118. M. Brower, M. Tennis, E. Denzler, M. Kaplan (1993) Powering the midwest renewable electricity for the economy and the environment. A Report by the Union of Concerned Scientists.
119. J.G. Slootweg (2003) Wind Power: Modelling and Impact on Power System Dynamics. In: PhD Thesis, Delft University of Technology, Netherlands.
120. S. Heier (1998) Grid Integration of Wind Energy Conversion System. John Wiley & Sons Ltd., Chicester, UK.
121. R. Resnick, D. Halliday (1961) Physics For Students of Science and Engineering-Part-I. John Wiley & Sons, Inc.
122. A.H. Church (1957) Mechanical Vibrations. John Wiley & Sons, Inc.
123. B. Morrill (1957) Mechanical Vibrations. The Ronald Press Company, New York.
124. Working Group on Prime Mover and Energy Supply Models for System Dynamic Performance Studies (1991) Dynamic models for fossil fuelled steam units on power system studies. In: IEEE Trans. on Power Systems, Vol. 6, No. 2, pp.753 – 761.
125. IEEE Std. 421.5 (1992) IEEE Recommended Practice for Excitation System Models for Power System Stability Studies.
126. PSCAD/EMTDC Manual (1994) Manitoba HVDC Research Center, Canada.
127. D.S. Zinger, E. Muljadi (1997) Annualized wind energy improvement using variable speeds. In: IEEE Trans. on Industry Applications, Vol.33, No.6, pp.1444 – 1447.
128. R. Hoffmann, P. Mutschler (2000) The influence of control strategies on the energy capture of wind turbines. Conference Record of the 2000 IEEE Industry Applications Conference, Vol.2, pp.886 – 893. Rome.
129. D. Driankov, H. Hellendoorn, M. Reinfrank (1993) An introduction to fuzzy control. Springer-Verlag.
130. P.N. Paraskevopoulos (1996), Digital Control System, Prentice Hall Europe.
131. A02-0165E (2001) ABB STATCOM for flexibility in power systems. ABB Report.
132. S.M. Schoenung, W.V. Hassenzahl (2003) Long- vs. short-term energy storage technologies analysis-a life-cycle cost study-a study for the DOE energy storage-systems program. Sandia National Laboratories, Report No.SAND2003 – 2783.
133. D. Linden (1983) Handbook of Batteries and Fuel Cells. McGraw Hill.
134. C. Lotspeich (2002) A Comparative Assessment of Flow Battery Technologies. In: Proceedings 2002 EESAT Conference, San Francisco, California.

135. M. Arulepp (2003) Electrochemical characteristics of porous carbon materials and electrical double layer capacitors. Dissertationes Chimicae Universitatis Tartuensis vol. 38, Tartu University Press.
136. Battery University, Practical battery knowledge for engineers, educators, students and battery users alike.
http://www.batteryuniversity.com. Cited 21 Feb 2008.
137. ELNA America, Inc., Uses of an Electric Double Layer Capacitor.
http://www.elna.co.jp/en, http://www.elna-america.com. Cited 15 Feb 2008.
138. R. Kottenstette, J. Cotrell (2003) Hydrogen storage turbine towers. National Renewable Energy Laboratory (NREL) report, NREL/TP-500-34656.
139. Docket No. RM05-4-001, Order No. 661-A (2005) Interconnection for wind energy. Federal Energy Regulatory Commission (FERC) report, United States of America.
140. J. Ribrant, L.M. Bertling (2007) Survey of failures in wind power systems with focus on Swedish wind power plants during 1997-2005. In: IEEE Trans. on Energy Conversion, Vol. 22, No. 1, pp.167 – 173.
141. J. Hirose, T. Isagawa (1997) A high-purity hydrogen and oxygen generator (HHOG) for chemical industry. In: Technical document of Shinkou Pantetuku, Vol. 40, No.2, pp. 48 – 56, in Japanese.

Index

FSC
www.fsc.org
MIX
Papier | Fördert gute Waldnutzung
FSC® C083411

Kolibri 360 GmbH
Ferdinand-Jühlke-Straße 7
99095 Erfurt, Deutschland
produktsicherheit@kolibri360.de